Contrails over *the* Mojave

The Golden Age of Jet Flight Testing at Edwards Air Force Base

George J. Marrett

NAVAL INSTITUTE PRESS
Annapolis, Maryland

Naval Institute Press
291 Wood Road
Annapolis, MD 21402

Library of Congress Cataloging-in-Publication Data

Marrett, George J.
The golden age of jet flight testing at Edwards Air Force Base / George J. Marrett.
p. cm.
Includes bibliographical references and index.
ISBN 978-1-59114-511-0 (alk. paper)
1. Airplanes, Military—California—Flight testing. 2. Edwards Air Force Base (Calif.)—History—20th century. 3. Airplanes—California—Flight testing—History—20th century. 4. Aeronautics—Research—United States—History—20th century. 5. Aeronautics, Military—Research—United States—History—20th century. I. Title.
TL671.7.M33 2008
623.74'60724—dc22

2007045302

Printed in the United States of America on acid-free paper ♾

14 13 12 11 10 09 08 9 8 7 6 5 4 3 2

First printing

All photographs in this book are from the U.S. Air Force, with the exception of photo 1.
Book design: David Alcorn, Alcorn Publication Design

To my wife, Jan,
who for fifty years has been the
wind beneath my wings

The skies

over the Mojave Desert

are the same as they were years ago,

blazingly hot summers and forbidding winds of winter

The contrails that marked the swift passage of these test pilots

toward death and destruction have been swept away

The memories of those days are now marked

by these words and so will forever be

engraved across

the sky

Contents

Foreword

My long career as a military pilot, from flying in combat in World War II and Vietnam to being in the business of flight testing for nine years, may qualify me to comment knowledgeably on George Marrett's latest book.

George uses the "golden age of jet flight testing" as the theme that pervades the stories of test pilots he writes about. Indeed, it was a time when experimental aircraft were entering the age of supersonic flight. Speed and altitude records were waiting to be broken, and the men who fell in love with flying during their boyhoods were there to accept the challenges and to push the envelope.

He writes eloquently about these aviators, bringing them to life with descriptions of them, their families, their motivations, and the drive that made them ready, willing, and able to voluntarily be among those who challenged the unknown. These men received both rewards for achieving success and the personal satisfaction of contributing to the utility and flying characteristics of the aircrafts they were testing. They knew that the primary role of the test pilot was not only self-satisfaction in a job well done but the enormous responsibility of ensuring that the best possible aircraft were delivered to the men and women who would fly those aircraft into combat in defense of our country.

One side of the coin was golden, but the other side was tarnished by the frequent accidents that cost the lives of test pilots. They were Marrett's friends and colleagues, and they were my friends too, since I had served at the Flight Test Center at Edwards Air Force Base in California in the years just before his service there.

George goes into great detail in describing the cause of each accident, whether it was the need for a split-second response by the pilot or a mechanical malfunction. He does this with respect and dignity for the pilot who lost his life doing what he loved, working hard to make sure his aircraft would be the best possible machine it could be.

On a lighter side, we fighter pilots, destined to test advanced aircraft that would introduce us to high G-force stress for extended periods, were required to undergo more than routine physical exams. Some of the testing was a bit embarrassing. Think: urine, how long can I hold it? George injects a bit of fighter pilot humor at just the right times. I laughed as his vivid accounts brought some of those occasions in my own life back to me.

When I read this book, as well as the others George has written, I felt that I was listening to a historian. When discussing an aircraft, he provides precise data on the aircraft performance, including speed, altitude, and range. He continues with exact information on the kinds of weapons it could carry and other features or qualities of that specific aircraft. I like that because, whether you're a pilot or an aviation enthusiast, I believe the mind's eye can better perceive a particular story as it unfolds if you know more about the aircraft.

I suspect George will be pleased when I say that I completely agree with his assessment of the F-111A and the F-4. He did not like either one. The F-111A was a mistake promulgated by Secretary of Defense Robert McNamara. It was to be the single aircraft for both the U.S. Navy and the U.S. Air Force. That was a monumental blunder. I flew the F-111A only once, but it turned out to be a dismal failure for the mission it was designed to accomplish. George details it well. I flew all the new Century Series fighters from the F-100 through the F-107 and flew the F-105 in combat in Vietnam. Years later I returned to Edwards Air Force Base as its commander and had a chance to check out the F-4. On my first flight, after a few routine maneuvers, I tried a dive-bombing run. On the pullout from the dive, the G-forces continued to increase and I had to push forward on the control stick to prevent an increase that could lead to a high-speed stall or loss of control. I flew the aircraft several times after that, but soon decided not to fly it anymore. The F-4 required special care or it could kill you. In combat, "special care" should be the last thing on the fighter pilot's mind. George draws some interesting conclusions after he performed an intensive test program on the F-4 to investigate some of its undesirable handling characteristics. They make for interesting reading.

The golden age of flight testing did come to an end. Rockets into space, earth orbits, and trips to the moon spelled their doom. The technological age assaulted us as analog electronics turned into digital and product reliability improved dramatically. Thanks in part to the computer age and sophisticated flight simulators, flight testing continues with far less loss of life than in years past. After he left the Air Force, George himself made a significant contribution to the golden age as a test pilot for the Hughes Aircraft Company—one of the great companies that contributed tremendously to the technical improvement of both civil and military aviation.

This book may speak about aircraft, but it is primarily about people, their lives, their families, their hopes and dreams, and the courage with which they faced the possibility of death on any given day. It is a must read—anytime—for any aviation enthusiast, but it will be equally enjoyable to those who have ever taken the safety of a commercial aircraft for granted.

Maj. Gen. Robert M. White, USAF (Ret.)
Former Commander
Air Force Flight Test Center
Edwards Air Force Base, California

Acknowledgments

Special thanks and recognition go to my wife Jan for her support of my writing and for her editing of this book. She has been my copilot for fifty years and, as she did with my other three books, has provided love, encouragement, and helpful suggestions.

To my older son Randy, professor of geology at the University of Texas, Austin, thanks for encouraging me to write about my aviation experiences and to record my flying history for our family.

To my younger son Scott, who was born at Edwards Air Force Base while I was a student at the Air Force Test Pilot School, thanks for reading and reviewing these stories. Scott is also a storyteller who greatly enjoys a well-told tale.

I am deeply grateful to my father, George Rice Marrett, who taught me the value of hard work and encouraged me to be honest and faithful to the U.S. Air Force, where I was employed for eleven years. He worked for just one firm, National Cash Register Company, for all his forty-nine years of employment.

Love to my mother, Julia Etta (Rachuy) Marrett, who at age ninety-seven still has a wonderful memory and sense of humor. Thanks, Mom, for encouraging me to write and for having faith in me.

Thanks to Richard Russell, director of the Naval Institute Press and my initial acquisitions editor for *Contrails over the Mojave,* for selecting this manuscript about flight testing at Edwards Air Force Base. It's ironic to me that a Navy publisher showed such great interest in an Air Force story. Thanks also to Susan Brook and Susan Corrado of the Naval Institute Press, who pushed my manuscript through the publishing process, and the outstanding work by my copy editor John Raymond and his assistant Katy Meigs.

I am particularly indebted to Dr. James O. "Jim" Young, chief historian of Edwards Air Force Base, and to his staff. Jim edited and fact checked my manuscript, provided photographs, wrote a cover quote, and provided me with a lot of support and encouragement during the twelve years it took to write this manuscript. Most of the material on the history of Edwards Air Force Base in chapter 1 is taken from a history of Edwards that Jim wrote, as is some of the material on Pancho Barnes in chapter 7, and they are used here with his permission. Long ago Jim encouraged me to document my years of flight testing at Edwards in book form as a permanent reminder to future generations of the speed and altitude records set by my generation. Though his name is not on the dust jacket, I think of him as my coauthor.

A special thanks goes to Maj. Gen. Robert M. "Bob" White, USAF (Ret.), for writing the foreword to this book. General White had the tour at Edwards of which we all dreamed about. He attended the Air Force Test Pilot School, graduating in January 1955, and staying at Edwards as a fighter test pilot. After the X-15's 57,000-pound-thrust YLR-99 engine was installed, he flew it to a speed of 2,275 mph in February 1961, setting an unofficial world speed record. He broke more speed records in succeeding months before reaching a speed of 4,093 mph on 9 November 1961, the first person to fly a winged craft six times faster than the speed of sound. Eight months later he took the X-15 to a record-setting altitude of 314,750 feet. Flying at this altitude qualified him for astronaut wings. Later, White was promoted to general and in the early 1970s became the Air Force Flight Test Center commander. Bob White had the ultimate test pilot career.

I want to thank the Air Force for permission to use material from General White's biography.

Appreciation also goes to Dr. Richard P. "Dick" Hallion, former chief historian at Edwards Air Force Base and retired chief historian of the U.S. Air Force in Washington, D.C. Dick had faith that I could transition from a pilot in the cockpit to an author on a typewriter and permanently record the historic flight test events that took place over the Mojave Desert during the mid-1960s.

I would like to recognize Fred Johnsen, former historian at Edwards and now the chief of public affairs at the NASA Dryden Research Center, for his encouragement through the long years it took to finish this manuscript. Fred has written on every aircraft from the A-1 Skyraider to the B-24 Liberator and was always available to answer questions.

Recognition must go to Paula Smith, executive director of the Society of Experimental Test Pilots (SETP), and her staff for assisting me in researching information on deceased members and encouraging me to write about test pilots.

The "Cuban Missile Crisis" and "Eye Patch" stories in chapter 5 originally appeared in *Testing Death: Hughes Aircraft Test Pilots and Cold War Weaponry* (Westport, Conn.: Praeger Security International, 2006). They are reprinted here with permission.

Earlier versions or portions of several chapters originally appeared in *Air & Space* (chapter 7), *Wings* (chapters 4, 5, 6, and 9), *Flight Journal* (chapters

4 and 10), *Aerospace Testing International* (chapter 1), and *Sabre Jet Classics* (chapter 3). I thank the editors—George Larson, Mike Machat, Roger Post, Christopher Hounsfield, and Larry Davis, respectively—for publishing them and for giving me a start in aviation writing.

Special recognition goes to Obbie Atkinson and Al Schade, two of the founders of the Estrella Warbird Museum in Paso Robles, California. Both are dedicated aviators and World War II military veterans who have encouraged me to write aviation history and who support my present flying in a 1945 Stinson L-5E Sentinel.

I especially want to thank the following individuals, some of whom connected me with other people who had information on Edwards Air Force Base, some of whom kindly agreed to be interviewed for this publication, and all of whom have supported my writing and publishing: Joe Benson, Walter J. Boyne, Pat H. Broeske, Margaret Cook, Marshall Cook, Jim Easton, Ed Gillespie, Robert J. Gilliland, Katherine Huit, Jesse P. Jacobs, Penny Jessup, Paul H. Kennard, Don S. Lopez, William R. Lummis, Ray Puffer, Jack Real, R. Kenneth Richardson, Barrett Tillman, and Darrel Whitcomb.

Last, I would like to recognize my four grandchildren: Cali Marrett, nineteen; Casey Marrett, seventeen; Tyler Marrett, fifteen; and Zachary Marrett, thirteen. Our grandchildren are our hope and dream for the future; they fill us with joy every single day. I pray that my writing will encourage them to someday put their life experiences in print.

Introduction

To be Born in Freedom Is an Accident
To Live in Freedom Is a Privilege
To Die for Freedom Is a Responsibility

They gave everything.

The simplicity of these three words explains the dedication of more than thirty of my test pilot friends who never returned from their last flight. *Contrails over the Mojave* is a story about these men, their love of flying, the world speed and altitude records they set, and the sacrifices they made for our country. To permanently record that illustrious history and honor their memory is the purpose of this book.

The book is a collection of stories that covers a period of about twenty years from the end of World War II through the Cuban Missile Crisis and until I left Edwards Air Force Base in 1968 to fly a year of combat duty during the Vietnam War. The stories occur during a time when world speed and altitude records were set over the Mojave Desert, a period historians have termed the golden age of jet flight testing.

My stories are about test pilots I knew and flew with but did not get a chance to say good-bye to. You probably don't know the names of my friends. Most of these pilots are not household names, as they were known only within the test pilot community. All were well respected by their fellow pilots, dedicated to their profession, and lost in accidents in the sky. Even after all these years, their deaths leave an emptiness in the hearts of the people who knew and loved them.

When I first started collecting my thoughts, I had no inkling this book was in the offing. Newspaper clippings and photographs I had saved were many years old. Initially, I spent many hours by myself remembering the thrill and excitement I had shared in the sky with my fellow test pilots. The names of lost friends seemed to go on and on.

One was my classmate in 1964 at the Air Force Test Pilot School (TPS). Two were in the class in front of me; another was in the class behind. Three were instructors at the school.

One was a pilot I worked for at the Edwards Air Force Flight Test Center (AFFTC) Flight Test Operations Fighter Branch after I graduated from Test Pilot School. Two were competitors of mine on the handball court in the base gym. Three were bomber/transport test pilots who let me, a fighter test pilot, fly in the cockpit of a large aircraft and get some flying time using a yoke instead of a control stick. For several, I flew in close formation as a safety chase pilot in a T-38A Talon, F-4C Phantom, or F-104A Starfighter, acting as an extra set of eyes, making sure the sky ahead was clear of other aircraft and that their aircraft appeared to be visually sound.

For some, I attended their memorial services and stood outside the base chapel at Edwards Air Force Base, California, watching a missing-man formation flyby. For others, I flew in the formation myself.

In every case we shared exciting moments in the cockpit and the sky. Most of what I learned about flight testing came through conversations and discussions with them and with those who survived about test techniques and procedures. It is my belief that they respected my thoughts and ideas as much as I did theirs.

Still, after more than forty years, I remember the joy and thrills we shared in the sky together, pushing the boundaries of aircraft performance and capability. Setting world speed and altitude records was a common occurrence. Our lives were full of hope and humor—we thought our test flying days would last forever.

The test pilots lost in these tragic accidents would have wanted us to keep pushing, for we were all on the same team. We were blessed to be entrusted with such magnificent machines and privileged to be a part of one another's lives. I felt it was important that a tribute be made to the memory of their sacrifices and accomplishments.

Chapter 1

Don't Kill Yourself

During the late fall of 1967 I was making a grand tour of Europe. Although I was merely a captain in the U.S. Air Force, an eight-passenger T-39 Sabreliner jet usually reserved for generals was my private mode of transportation. I was flying in glorious style.

For the previous three years I had been a fighter test pilot assigned to the Air Force Flight Test Center (AFFTC) at Edwards Air Force Base in California, flying the F-4C Phantom, F-5A Freedom Fighter, F-104A Starfighter, and F-111A Aardvark. I wore an orange flying suit, a bright blue custom flight helmet, and high-speed ejection spurs as I accelerated to Mach 2 in my silver beauties and left contrails over the Mojave Desert.

I had recently completed several extremely dangerous test flights in the F-4C trying to determine why eight aircraft flown by Tactical Air Command (TAC) student pilots had crashed while maneuvering at low altitude (fortunately, none of the pilots died). Senior commanders in TAC were anxious to hear the results of my flight tests so I gave a detailed briefing to them at their headquarters at Langley Air Force Base in Virginia. Fearing that more F-4Cs would be lost and pilots killed, the officers asked me to give presentations to operational F-4C units all over the United States. Interest in my Phantom flight-test results also came from the headquarters of the U.S. Air Force in Europe. A request came for me to brief the European headquarters staff in Wiesbaden, Germany, and all the F-4C squadrons in England, Germany, and the gunnery range at Wheelus Air Base in Libya. So a T-39 was made available to transport a junior captain in high fashion.

The Air Force version of the F-4 Phantom was an outgrowth of the U.S. Navy F4H. The McDonnell Aircraft Corporation F4H was a twin-engine, 40,000-pound interceptor used for fleet air defense. Armed with air-to-air AIM-7 Sparrow and AIM-9 Sidewinder missiles, the Phantom could orbit 250 miles from the carrier for two hours, preventing enemy aircraft from attacking the fleet. The Phantom was the first naval fighter to dispense with cannon armament and had a weapon system that could identify, intercept, and destroy any target that came into range of its radar without having to rely on ground control.

It was also the first fighter originally designed solely as a carrier-based fighter to be ordered by the Air Force. The F-4 was built in great numbers. It was the second-most produced U.S. jet fighter, outnumbered only by North American Aviation's F-86 Sabrejet. Total U.S. production was over five thousand fighters. The Phantom was in continuous production for twenty years (from 1959 until 1979). During the Vietnam War, seventy-two Phantoms came off the production line every month.

When the Air Force accepted the F-4, they were told to make minimum modifications to the aircraft. Since the Phantom would be landed on a runway instead of on a carrier, the Air Force installed larger tires and antiskid brakes. In addition, the Air Force wanted to fly the F-4C in combat in Vietnam both as an air-to-air fighter and as an air-to-ground bomber. Therefore, all nine of the underwing weapon stations would be used. For bombing missions the Phantom would be loaded with fuel tanks and armed with bombs, napalm, cluster bomb units (CBU), and rockets. All the external fuel and ordnance would be attached to the underside of the wings and fuselage.

Instructors were training student pilots in combat crew training squadrons all over the United States at a fast pace. The war in Vietnam was ramping up and there was a great need for combat-ready pilots. In preparation for flying training missions, F-4C aircraft were loaded with small practice smoke bombs, LAU-19 rocket pods (each pod held nineteen rockets), and two 370-gallon underwing fuel tanks. This loading greatly changed the center of gravity (CG) of the plane from the Navy version and, ultimately, the flying characteristics of the beast. The ordnance was fired or dropped on simulated targets in gunnery ranges. After takeoff, with a full load of fuel, the aircraft was being flown near the aft CG limit. This meant the aircraft was very difficult to handle with extremely light flight controls. All eight planes crashed on test ranges within twenty-five minutes of takeoff.

TAC commanders asked the Flight Test Center to run a test program and determine why these F-4 Phantoms had crashed. I was given the task of writing a test plan, getting the plan approved by the center commander, flying the missions, and finally writing a test report.

I prepared my briefing material and showed up at the office of Center Commander Maj. Gen. Hugh B. Manson one morning. As I waited outside the general's office I reviewed the questions he might ask. Did we have the correct instrumentation on board? Was the program properly funded? Was the Phantom in good mechanical condition? Did I feel trained enough to perform the maneuvers?

I gave my pitch in just a few minutes. He nodded his head a couple of times but did not interrupt me with questions. When I finished I paused for a response. He looked out the window for a few moments. Then he turned to me and talked about the recent accidents and deaths of several test pilots under his command. I listened intently, trying to sense the question that would likely follow.

Finally, he looked me straight in the eye and said, "Don't kill yourself!"

Death had been a frequent visitor to the Mojave Desert. A parched and forbidding wilderness to those who merely passed through it, the northwestern Mojave Desert was a land of kangaroo rats, coyotes, and Joshua trees. It was a harsh land of sometimes stunning contrasts—a land of bone-chilling nights and griddle-hot days, of violent dust storms and mesmerizing sunsets.

According to a history of Edwards Air Force Base written by its chief historian, James "Jim" Young, until the Southern Pacific Railroad arrived in 1876, it was a land peopled largely by prospectors drifting aimlessly and endlessly in pursuit of elusive mineral wealth. Some died from lack of water and exposure to the sun. It was a sign of times to come.

In 1882, the Santa Fe Railway ran a line westward out of Barstow, California, toward the town of Mojave. A water stop was located at the edge of an immense dry lake roughly twenty miles southeast of Mojave. The water stop was known simply as Rod, and the lakebed was then called Rodriguez Dry Lake. By the early years of the new century, Rodriguez had been anglicized into Rodgers, which was, in turn, ultimately shortened to Rogers. First formed in the late Pleistocene epoch, Rogers Dry Lake has an extremely flat, smooth, and concretelike surface. A playa—or pluvial lake—it spreads out over forty-four square miles, making it the largest geological formation of its kind in the world. Its parched clay and silt surface undergoes a cycle of renewal every year as water from winter rains is swept back and forth by desert winds, smoothing it out to an almost glasslike flatness.

In 1910, Clifford and Effie Corum, along with Cliff's brother Ralph, settled at the edge of this lakebed. In addition to raising alfalfa and turkeys, the enterprising Corums got others to homestead in the area for a fee of $1 per acre. As those settlers moved in, the Corum brothers got contracts for drilling their water wells and clearing their land. They also opened a general store and post office. Their request to have the post office named Corum, after themselves, was disallowed because of potential confusion with the already existing Coram, California. So they simply reversed the spelling of their name and came up with Muroc. The name stuck for nearly forty years. As had happened so often in the history of the American frontier, the Corums and their neighbors had the wherewithal to make their living from an inhospitable land. They built institutions such as a public school and the Muroc Mercantile Company and post office that, under various names, would serve as a focal point for community activities up until the late 1940s. Perhaps even more important, the Corums and their fellow settlers bequeathed an enduring legacy—a pioneering spirit that characterized the careers of many of those who followed as they confronted yet another kind of frontier, a frontier of aviation.

While the early homesteaders may have thought of it as a wasteland, in the eyes of a visionary airman such as Lt. Col. Henry H. "Hap" Arnold in the early 1930s, it was a one-of-a-kind natural aerodrome—one that could be employed at virtually no cost to the taxpayer. And, by the early 1950s, after having already witnessed countless milestone flights in the clear blue sky above its sprawling expanse, flight test pioneer and Air Force Flight Test Center Commander Brig. Gen. Albert Boyd would think of it as nothing less than "God's gift to the U.S. Air Force."

Muroc Bombing and Gunnery Range, within the confines of present-day Edwards Air Force Base, was first established by Colonel Arnold as a remote training site for his March Field squadrons in September 1933. It continued to serve in that capacity until July 1942 when it was activated as a separate post and designated Muroc Army Air Base. Throughout the war years, the primary mission at Muroc was to provide final combat training for aircrews just before overseas deployment. In the spring of 1942, however, the immense volume of flight testing already being conducted at Wright Field in Dayton, Ohio, was one of the factors driving a search for a new site where a top-secret airplane could undergo tests. The aircraft's highly classified nature compelled program officials to find an isolated site away from prying eyes. The urgent need to complete the program without delay dictated a location with good, year-round flying weather. And the risks inherent in the radical new technology to be demonstrated on the aircraft dictated a spacious landing field. After examining a number of locations around the country, they selected a site along the north shore of the enormous, flat surface of Rogers Dry Lake, about six miles away from the training base at Muroc.

The aircraft was America's first jet, the Bell XP-59A Airacomet. On October 1, 1942, Bell test pilot Bob Stanley lifted the wheels of the jet off the lakebed for the airplane's first flight, and the turbojet revolution belatedly got under way in this country. Both the British and Germans had already flown jet aircraft. Like all of the early turbojets, the Airacomet was woefully underpowered and it required extremely long takeoff rolls. This, plus the fact that the new turbojet engines had a nasty habit of flaming out, confirmed the wisdom of selecting the vast forty-four-square-mile expanse of the lake bed to serve as a landing field. In years to come, it would become a welcome haven to countless pilots in distress.

The XP-59A introduced flight testing to the Mojave—but the testing that took place at the small Materiel Division test site (now known as North Base) bore little resemblance to what has evolved since. There were no aircraft flying chase for the Airacomet's first flights. There was no telemetry or mission control center, just a portable two-way radio and a voice recorder on the ramp outside the hangar. The most important sources of data during those first few flights were the pilot's kneepad notes and, perhaps the most critical instrumentation, the seat of his pants—not too scientific but, by later standards, relatively inexpensive and affording a means of real-time data acquisition that yielded immediate analyses of any problems. Ultimately, an instrumentation panel, filmed by a camera activated by the pilot, was installed, and the aircraft was instrumented to collect data on about twenty parameters. Much of the instrumentation, however, remained primitive. Control-stick forces, for example, were measured with a modified fish scale. It was an age when old-fashioned common sense and improvisation still ruled supreme—and slide-rule data reduction and analysis took days of painstaking manual effort.

As with virtually all of the test programs conducted during the war years, most of the flight-test work on the P-59 was conducted by civilian contractors. Although Army Air Forces (AAF) pilots flew the aircraft from time to time, and flight test engineers from Wright Field reviewed the data, the formal preliminary military test and evaluation program did not commence until the fall of 1943, a year after the first flight. Designed to validate the contractor's reports, this preliminary evaluation consisted of a very limited number of flights and was essentially completed within a month. Formal operational suitability and accelerated service tests did not get under way until 1944, well after the AAF had decided that the airplane would not be suitable for combat operations and would, instead, be relegated to a training role.

The P-59s were tested at Muroc from October 1942 through February 1944 without a single accident. Though the aircraft did not prove to be combat worthy, the successful conduct of its test program, combined with the success of the Lockheed XP-80 program that followed it in early 1944, sealed the destiny of the remote high desert installation. Muroc would thenceforth become synonymous with the cutting edge of the turbojet revolution in the United States. All of America's early jets—both Air Force and Navy—underwent testing at Muroc, and the successful conduct of these programs attracted a new type of research activity to the base in late 1946, after the end of World War II.

The rocket-powered Bell XS-1 was the first in a long series of experimental airplanes that were designed to prove or disprove aeronautical concepts—

to probe the most challenging unknowns of flight and solve their mysteries. On October 14, 1947, Capt. Charles E. "Chuck" Yeager became the first man to exceed the speed of sound in this small bullet-shaped airplane. He had been thrown off a horse and had broken a rib the night before.

By later standards, the XS-1 flight test team that accomplished this feat was extraordinarily small, never numbering more than about fifteen to twenty Air Force, National Advisory Committee for Aeronautics (NACA, later National Aeronautics and Space Administration, NASA), and contractor personnel. The seat of the pilot's pants remained important, but onboard instrumentation now covered about thirty parameters, the use of radar for tracking was introduced, and there was even some very preliminary, experimental use of telemetry. There was no formal safety review process, however, not even for the most hazardous programs, such as the X-1's envelope expansion tests. Beyond the simple maxim "Don't go too far, too fast," Captain Yeager and project engineer Capt. Jack Ridley set their own limits. They sat down just before a mission and decided how far and how fast, and then Yeager went up and flew the designed profile.

With the X-1 flights at Muroc, testing began to assume two distinct identities. Highly experimental research programs—such as with the X-3, X-4, X-5, and the XF-92A—were typically flown in conjunction with NACA and conducted in a very methodical fashion to answer largely theoretical questions. The bulk of the testing focused on highly accelerated Air Force and contractor evaluations of the capabilities of new prototype aircraft and systems proposed for the operational inventory. Not surprisingly, the rather informal approach to safety that prevailed during the late 1940s—and even into the 1950s—was one of the factors contributing to a horrendous accident rate. In addition, the group of test pilots at Muroc remained very small, and thus they averaged nearly a hundred flying hours per month, year after year. They flew an enormous number of types and models of aircraft, each with its own cockpit and instrument panel configuration. Yeager once flew twenty-seven different types of airplanes within a single one-month period.

Many aviation historians believe that the twenty years from 1927, when Charles Lindbergh flew solo across the Atlantic Ocean, to 1947, when Chuck Yeager first flew supersonic in the Bell XS-1 over Edwards, was the golden age of propeller aircraft. While Lindbergh cruised at 100 mph on his record-setting flight, Yeager blasted over the Mojave Desert nearly seven times faster, leaving a contrail in the bright blue sky. A new era started that fall day, an era of twenty years, from 1948 to 1968. During that period military and civilian test pilots set a multitude of world speed and altitude records over Edwards. The extreme pressure of the Cold War with the Soviet Union caused the Air

Force to fund all types of aeronautical research with the hope that United States could maintain air supremacy. The price of developing these unique aerial vehicles, to set speed and altitude records and to keep a lead on the Soviets, was high, both financially and in the lives of a generation of test pilots.

The year 1948 was particularly tragic, with at least thirteen fatalities recorded. One of them was Capt. Glen W. Edwards, who died in the crash of a Northrop YB-49 flying wing bomber. In December 1949, the base was renamed in his honor. By that time it had long since become the de facto center of American flight research, and, on June 25, 1951, this fact was finally given official recognition when it was designated as the U. S. Air Force Flight Test Center (AFFTC). That same year, the USAF Test Pilot School (TPS) moved to Edwards from Wright Field. The curriculum focused on the traditional field of performance testing and the relatively new field of stability and control that had suddenly assumed critical importance with the dramatic increases in speed offered by the new turbojets. Increasingly, as the aircraft and their onboard systems became ever more complex throughout the 1950s, those selected for admission to the school had to be far more than outstanding pilots; they also had to have the technical backgrounds that would enable them to thoroughly understand all of the systems they would be evaluating and permit them to translate their experiences into the very precise jargon of the engineer.

By any standard, the 1950s was a remarkable period in the history of aviation, and there was no better evidence of this than what transpired at Edwards where, if a concept seemed feasible—or even just desirable—it was evaluated in the skies above the sprawling three-hundred-thousand-acre base. When NACA test pilot Scott Crossfield first arrived on base in 1950, he found it "hard to believe that this primeval environment was the center of aviation's most advanced flying." He likened it to an Indianapolis 500 of the air. But it was more than that. It was "an Indianapolis without rules," because the test pilots at Edwards "lived with the feeling that everything they were doing was something that probably had never been attempted or even thought of before." Crossfield became most closely identified with a series of experimental aircraft that had been launched with the X-1, and certainly the most publicized activity at Edwards throughout the 1950s continued to be in the realm of flight research where the limits of time, space, and the imagination were dramatically expanded.

The experimental rocket planes continued to expand the boundaries of the high-speed and stratospheric frontiers. As the decade opened, the first-generation X-1's Mach 1.45 (957 mph) and 71,902 feet represented the edge of the envelope. The Douglas D-558-II Skyrocket soon surpassed these records.

In 1951, Douglas test pilot Bill Bridgeman flew it to a top speed of Mach 1.88 (1,180 mph) and a peak altitude of 74,494 feet. Then, in 1953, Marine test pilot Lt. Col. Marion Carl flew it to an altitude of 83,235 feet, and on November 20 of that year, Crossfield became the first man to reach Mach 2, as he piloted the Skyrocket to a speed of Mach 2.005 (1,291 mph). Less than a month later, Maj. Chuck Yeager obliterated this record as he piloted the second-generation Bell X-1A to a top speed of Mach 2.44 (1,650 mph), and nine months later, Maj. Arthur "Kit" Murray flew the same airplane to a new altitude record of 90,440 feet.

These records stood for less than three years. In September 1956 Capt. Iven Kincheloe became the first man to soar above 100,000 feet, as he piloted the Bell X-2 to a then-remarkable altitude of 126,200 feet. The following year he was killed in a Lockheed F-104 Starfighter. Flying the Bell X-2 just a few weeks later, on September 27, Capt. Milburn "Mel" Apt became the first man to exceed Mach 3, as he accelerated to a speed of Mach 3.2 (2,094 mph). His moment of glory was tragically brief, however. Just seconds after attaining top speed, the X-2 tumbled violently out of control and Apt was unable to recover it. Apt died in the crash. With the loss of the second X-2—the first had crashed earlier—the search for many of the answers to the riddles of high-Mach flight had to be postponed until the arrival of the most ambitious of all the rocket planes—the truly awesome North American Aviation X-15. The X-15 would continue to break world speed and altitude records and cause death and destruction after I arrived on the windswept Mojave Desert in late 1963.

Chapter 2

Chasing the Whistle

A blackout was in progress and all lights inside and outside our house were extinguished. Only the one-inch-square radio dial was illuminated. In early 1942, my parents and I were sitting next to a console radio listening to such programs as *Amos 'n' Andy*, the *Jack Benny Program*, and *Fibber McGee and Molly*. Frightening news reports about air battles in World War II occasionally interrupted these programs. I was seven, and these were some of my earliest memories.

Our family lived in Grand Island, Nebraska, about 130 miles west of Omaha, in the heart of the Midwest, more than a thousand miles from either coast. However, many people feared that the Germans or the Japanese would overfly the United States and bomb the Boeing B-17 Flying Fortress bomber base just outside our town. Because the lights around our houses could help the enemy bomber pilots find the base, the town had to be completely dark. Every couple of months sirens would signal the start of the practice blackout. Then an air raid warden, wearing a doughboy-type helmet, knocked loudly on our door, interrupting the radio program. He checked us for compliance with the rules. Watching all of this activity while sitting in the dark was very scary for a small boy.

During the day, I saw and heard formations of B-17s fly over our house. It was the loudest noise and rumble I had ever heard in my life; our house shook. The sound of those huge roaring radial engines is still locked in my memory over sixty-five years later.

Bob Preston, my best friend, lived just a block away. At age eight, he was a year older than I and chose the games we played. He taught me to play a game called "Pilot." We each wore a leather flying helmet and goggles as we pretended to be airplanes chasing each other across the sky in our front yard. We ran around trees and buildings, making the sound of a fighter plane chasing after the dreaded enemy that scared me during the occasional blackouts. When Bob played the part of the enemy, I easily chased him down and scored a kill. I could run faster and won every time.

When we played bomber pilot, Bob was always the pilot and always made me be his copilot. We flew in the same plane and I took orders from Bob. According to Bob, you needed a silver metal whistle attached to a chain and pinned to your shirt to be a pilot. Without a whistle a person was destined to be no more than a copilot. I envied Bob and his whistle and promised myself that someday I would get a whistle and advance into the lofty ranks of the pilots. I never asked Bob why a whistle was required or what purpose it served. It was just a requirement—that was enough for a young boy.

That little silver whistle became very symbolic during my later flying career. No matter how far I advanced in flying, a symbolic new whistle was always required. There was always a higher achievement over the horizon. Initially, I was not considered, by instructors and other qualified pilots, as a pilot in the Air Force until I soloed, then I was not a pilot until I flew jets. After that, I was not a pilot until I got my silver wings in Air Force flying school. Then I was not a pilot until I became combat ready in a fighter squadron. I was always chasing the whistle.

The same was true in the test pilot field. Initially, I was not considered a test pilot until I graduated from test pilot school, then I was not a test pilot until I had published a test report and made a presentation on the results. Then I was not a test pilot until I had led a test team on a major program . . . chasing the whistle again.

I started building wooden model airplanes at age seven, during the war. At that time a kit could be purchased for five cents and included balsa wood, paper for covering the surfaces, a small propeller, a rubber band, and assembly instructions. My first completed plane was a Stinson L-5 Sentinel, a two-place, high-wing, tail-wheeled craft used as an artillery spotter. It was a beauty and propelled my mind for many hours through the Nebraska skies. To this day, that model plane remains perfect in my memory.

To support the military effort during World War II, many goods were rationed to civilians. Gas rationing had a large effect on people's lives. Fortunately, my father worked as a repairman for the National Cash Register Company and his services were vital to keeping the local businesses operating. He was always able to obtain gas, so we could make short family trips.

White sugar was also in short supply. We used brown sugar as a substitute, but I thought it looked and tasted terrible. Beef could be purchased only if you had the small red plastic tokens distributed by the government. As a substitute, my father raised rabbits in our side yard and sold the meat and fur to neighbors. My job was to feed the rabbits while my father was working. I became quite attached to the bunnies and hated to see them grow up. Since I helped my father slaughter and skin the adult rabbits, I knew what future was in store for them.

The government requested that civilians donate all the aluminum found in their homes and farms. Aluminum was used to build warplanes. A few blocks from our house was a huge pile of pots, pans, and other aluminum items that neighbors had given up. From a distance, I could see a beautiful large aluminum toy airplane wedged in the scrap metal. How I longed to own that plane. I have never seen a toy that matches my recollection of that plane.

I joined the Cub Scouts when I was nine years old. Several boys my age met in our basement and my mother was the den mother. My project involved collecting two thousand pounds of scrap newspaper to be given to the government in support of the war. The huge amount of paper I collected practically filled my father's one-car garage. By meeting the goal, I qualified for my first award, the General "Ike" Eisenhower Medal. I was very proud of my achievement; I had helped the war effort.

During the war our family visited friends in Columbus, Nebraska, about sixty miles away. This family had a boy my age, Don Lundin. Don's father made arrangements for him to get a flight in a Piper J-3 Cub, a plane very similar to the Stinson L-5 Sentinel model I had built. I was offered a chance to fly in the plane with Don. Since we were both small and only nine years old, Don and I were strapped together in the backseat of the plane. The seat was designed to hold one adult. We sat side by side, the control stick between our knees and a single seat belt holding both of us in our seat. Taxiing on a dirt strip, the light plane bounced around, throwing us against each other. When we got airborne, the bouncing did not stop, it intensified. I expected the flight to be smooth and the plane maneuverable like my body as I ran around my make-believe sky in our yard. As the pilot moved the control stick left and right to extend the ailerons and place the plane in a bank, the stick was pressed against either Don's knee or mine. We would immediately glance at each other with a child's amazement and delight. It didn't matter that I couldn't recognize anything on the ground or that I didn't know where we were or that the airplane was bouncing all over the sky—this was fun and exciting. Flying even looked easy. The hook was set. I decided to become a pilot someday.

My mother had first flown in an airplane in 1926. She was sixteen years old and living in Little Rock, a small town in northern Iowa. A barnstormer passed through town and gave a free ride to my mother and her older brother. She remembered the plane making dips in the air and that her brother's hat flew off in flight. She had a great time and the hat was later found in a farmer's field.

By late 1944, German prisoners of war (POWs) were transported from Europe and housed in a local elementary school in the center of our town. A tall barbed-wire fence circled an entire square block. One evening just before Christmas, my father drove our family downtown and parked on the street in front of the school. Under glaring spotlights, the German prisoners, dressed in khaki uniforms with a white "PW" stenciled on the front of their jackets, sat on bleachers facing the street. They sang Christmas carols in German. The enemy POWs did not look dangerous behind the fence, but I stayed in the backseat of our car and rolled my window down only a couple of inches.

Toward the latter part of the war, I got my first job. I delivered the *Omaha World Herald*, our morning newspaper, seven days a week. The newspaper cost five cents during the week and seven cents on Sunday. A stack of papers was delivered to my house early in the morning. My job was to fold each individual paper, organize the copies in a cloth fabric bag attached to my bicycle, and deliver them on my route.

Bicycling on the sidewalk, I tossed the paper, in what later would be known as Frisbee-style, onto the subscriber's porch. Each paper had to be placed on the porch in front of the door regardless of rain, snow, or wind. I was given two free papers of the *World Herald* as spares in case my toss was not accurate and the paper ended up on the roof. It was firmly planted in my mind to never miss the porch with my toss. I had found another use for the spare papers.

The local train station of the Union Pacific Railroad was only a few blocks from the end of my paper route. During the war trains discharged soldiers to be trained at the local air base and took aboard soldiers already trained and going off to war. I waited for the train to stop, jumped aboard, and walked down the aisle offering to trade a newspaper for a soldier's spare military cloth patch. The cloth patches were sewn on the soldier's sleeve, and indicated the rank, unit, or crew position in an aircraft. They had spares in their duffel bags. Soldiers were eager to learn the latest news about the war; it was an easy and profitable trade. In a short time, I had collected a complete set of Army Air Forces patches. My mother sewed the patches on a large cloth banner, which hung on my bedroom wall for years. I still treasure the banner and plan to donate it to an aviation museum.

The war ended just before my tenth birthday. As I bicycled around town with red-white-and-blue crepe paper threaded between the spokes of my bicycle, people were honking their car horns. Others were screaming out to one another with great excitement and joy. On the back of my bicycle, I had placed my sister's toy cloth monkey dressed as Tojo, the hated dictator of the Japanese. Tojo's neck was tied to a rope that was suspended from a long stick. I thought that Tojo should hang. Even I could tell that our country had suffered much during the previous five years of the war. Now the war was over.

After the war, at the age of twelve, I joined the Boy Scouts. Surplus used World War II Army equipment was now on the open market for civilians to purchase. My father bought an official military canteen, mess kit, cot, and bedroll for me to use on our Scout camping trips. A returning World War II soldier brought back a German metal army helmet and gave it to me. I wore the helmet on every campout. In many respects, our Scout troop looked more like a military unit than a Scout organization.

Camping in the cold Nebraska winters taught me to become organized. Without extensive planning and preparation, I would get very cold and hungry. The organizational ability I learned in the Boy Scouts became very helpful to me in later life, especially when I learned to fly many years later. Through the Boy Scouts, I learned self-reliance, teamwork, and independence. These life patterns remain with me to this day.

Membership in the Boy Scouts also strengthened my love of our country. As we traveled to campsites around the state, I observed the great outdoors of the central plains. I enjoyed being outside and learned to identify everything in nature.

My church sponsored our Scout troop and I began to develop a spiritual life. The values taught in the Scout movement could not, in my judgment, be improved on even now, sixty years later. The Scout Oath: "*On my honor I will do my best to do my duty to God and my Country and to obey the Scout Law. . . .*"

In retrospect, even though I was born during the Great Depression, grew up during World War II, and lived in a state with a harsh climate, it was a great time and place for a young man to come of age.

In high school my interests turned to mathematics and science. I especially liked chemistry, where I learned to make gunpowder. From watching World War II newsreels, I observed Army Air Forces P-47 Thunderbolts strafe German trains with their guns blazing. The boilers exploded and the trains crashed. Pilots were not needed in peacetime to destroy trains, so I gave up on my plan to be a pilot. I did enjoy blowing things up, though, so I thought about becoming an explosives expert.

On my own, taking a large metal bolt, I rotated a matching nut a few threads onto the bolt to construct a container. Then I filled the opening with homemade gunpowder, screwed another bolt to the other side of the nut, and cinched it tight. I had built my first weapon. I dropped this weapon off the roof and into the alley of a building downtown. It caused a large explosion and metal shrapnel flew in all directions. I was too scared to do it again.

Each weapon I constructed became larger than the last. Several times I started chemical fires in my parents' basement. They were not pleased with the odors and smoke that filtered into the main part of the house. Orders to use more care and good judgment went unheeded.

One day my mother read in the *Grand Island Independent*, our local newspaper, that a boy my age had been making pipe bombs. He had built the largest weapon he had ever made, lit it, and attempted to throw it. The bomb exploded in his hand. He was rushed to the hospital where his arm had to be amputated. My mother suggested I visit the hospital and learn from him how to make pipe bombs. When I arrived at the hospital and met him in his room, my interest in explosives took a sudden downturn. No more bombs, rockets, and explosives for me. I then realized why my mother had told me to meet the young man. I decided to become a scientist specializing in fundamental research.

My parents moved from Grand Island, Nebraska, to Sioux City, Iowa, after I graduated from high school in 1953. To qualify for in-state tuition, I enrolled at Iowa State College in Ames, Iowa, with a major in chemistry. The Korean War was just over and many veterans were also enrolling. Iowa State was a land-grant college supported by the federal government. As such, all undergraduate men were required to sign up for two years in the Reserve Officers' Training Corps (ROTC) in order to retain their draft deferment. I chose the Air Force Reserve Officer Training Corps (AFROTC).

Undergraduates in AFROTC were given a blue military blouse, pants, two long-sleeved dress shirts, a field cap, and a tie. Since military shoes, winter overcoat, and gloves were not issued, cadets could wear their own coat, shoes, and gloves. This was a license to dress as outrageously as possible and get away with it. Everyone deliberately dressed in a different color and style. We looked like George Washington's army crossing the Potomac.

During the winter of my second year in college, the weather in Iowa turned brutally cold and snowy. The temperature dropped to 15 degrees below zero and the snow crunched as I walked on it. I found out from another cadet that our AFROTC professor of air science, a full colonel and pilot, had scheduled a weekend pleasure flight to Florida in a Douglas C-47 Gooney Bird. Openings were available for cadets and I signed up immediately.

When twenty cadets arrived by bus at the civilian airport in Des Moines, Iowa, the plane was sitting on the ramp, its wings covered with ice. We boarded the plane, and it was towed into the hangar and the hangar doors were closed so that warm air would melt the ice. The pilot wanted all cadets strapped in their seats and ready to go. As soon as the ice melted, the plane was towed out, the engines were started, and the plane took off before more ice could accumulate.

Our first stop on that Friday afternoon was at Scott Air Force Base near St. Louis, Missouri. Snow could not be seen from the airplane window, and the temperature was still below freezing. Friday night we stayed at Craig Air Force Base in Selma, Alabama. It seemed surprisingly warm to me for wintertime, probably in the low 50s. Saturday morning we flew over the green terrain of Florida, landing twenty miles south of Miami Beach at Homestead Air Force Base. I was astounded. I had thought that when it snowed in the Midwest, it snowed across the entire United States. This was long before weather satellites and national weather forecasts.

That evening another cadet and I visited the Fontainebleau Hotel, a luxurious hotel located on Miami Beach. The hotel was about fifteen stories high, surrounded by palm trees and many swimming pools. It had a glass-walled cocktail bar that seemed to be a bar-within-a-bar-within-a-bar. I had never seen such splendor and luxury. Even at night, the temperature was in the balmy 70s. It dawned on me that aviation was a career field in which a person could escape the winter conditions I had grown up in. It made sense to apply for the two-year advanced Air Force ROTC program and become a flying officer. The Cold War had gradually heated up during the 1950s. Pilots were again needed to protect the peace, just as in World War II. My dream about becoming a pilot came back to life.

After the winter trip to Florida, I applied for advanced Air Force ROTC. My academic grades were satisfactory, and I easily passed the flight physical examination. All that was now required was to appear before an Air Force officers' board for an interview.

Since it was considered a badge of honor to wear the undergraduate uniform as shabby looking as possible, I now had to pay a lot of attention to my uniform to make a good impression on the board. For the first time in two years, I had my military blouse and pants dry-cleaned and pressed. My blue dress shirt was washed and ironed and my black dress shoes were shined. Next came a short haircut and close shave.

As I walked over to the armory building I reviewed why I wanted to become a military officer and pilot. I tried to think of the questions I might be asked and the answers I would give. Arriving half an hour before my scheduled interview, I sat on a long wooden bench outside the interview office and talked out loud to myself. I wanted to sound positive and self-assured. My voice seemed raspy, so I decided to drink a Coke and practice a little more.

In the 1950s soda came in a glass bottle with a metal cap that could be released with an opener attached to the Coke machine. I opened the Coke, walked back to the wooden bench, sat down, and started to drink. After a few swallows my voice sounded better, more like an officer's. When the Coke was empty, I stood up, planning to return the bottle to the rack. As I did so, I realized that I had inadvertently sat on a piece of pink bubble gum. The bubble gum was stuck to the back of my right pant leg and was strung from my pants to the wooden bench. I used my right thumb and forefinger to try to remove the gum from my pants, without success.

What should I do?

At that very moment, the door opened and an enlisted airman announced, "Cadet Marrett, the officers are ready to see you."

I did not have time to remove the gum. I entered the interview room and observed five Air Force officers sitting behind a long table. I marched into the room, stopped in front of the table at military attention, and saluted the officers. With my right hand held to my forehead, a long string of pink bubble gum extended from my finger to my rear end.

The officers were accustomed to pranks played by college students. However, this display of military bearing had taken pranks to a new level. Either I was one gutsy cadet or a real loser. The expressions on the officers' faces seemed to be mixed.

The senior officer asked me to define the word *discipline*. My mind went blank. As I spoke, struggling to make any sense at all, my voice began to change up and down in pitch as though I were fifteen years old. I thought to myself, I'm definitely not officer material.

As I walked back to my college dormitory, I thought about my short military career. I did not expect to be accepted into advanced AFROTC after making such a poor showing at the interview. To my surprise, a few days later I received a letter announcing my acceptance as an officer and a gentleman in the Air Force. I'm sure the officers on the review board never forgot the twenty-year-old bubble-gum cadet with his twenty-gum salute!

In 1956, during the summer between my junior and senior years in college, I attended a three-week Air Force ROTC summer camp at Fairchild Air Force Base in Washington state. I drove there with two other cadets, also from Iowa State College, in a 1953 black Mercury convertible. The cadet who owned the car let the other two of us take turns driving. He stripped to his shorts and lay down in the backseat to get a suntan. Crossing Nebraska, Wyoming, Montana, Idaho, and into Washington took three days of driving, and by the time we arrived his skin was bright red. Our cadet uniforms were made of cotton and starched extremely stiff. When the sunburned cadet tried to dress in his uniform, the pain was unbearable. He ended up in the base hospital for the entire three-week tour. One cadet down, two to go.

This was the first time I had been fully exposed to military life. The day began at 5 AM with physical training, followed by breakfast, marching, and more marching. Classes were conducted on military customs, the history of the U.S. Air Force, and survival training. After the evening meal, we had inspection. If a cadet passed, he could visit the cadet club and drink Olympia beer. If he failed, he marched off his demerits. The entire time, both day and night, was filled with cadet activities. With all this activity, I did not have an urge to have a bowel movement.

Fairchild Air Force Base was the home base of a Convair B-36 Peacemaker bomber wing. This aircraft was designed during the latter part of World War II, but it was not built in time to engage in combat. The aircraft had six pusher reciprocating engines and four jet engines. It was the largest aircraft in the Air Force inventory and was used by Strategic Air Command (SAC) as an offensive nuclear bomber. As cadets, we looked forward to our flight in the B-36. The flight was offered to cadets to encourage them to prepare themselves for USAF flight training.

The crew of the bomber we flew had jointly purchased a 1940s Crosley automobile. The Crosley was the smallest auto ever built by a U.S. manufacturer, and it fit in the B-36 bomb bay on training missions. When the crew landed somewhere other than their home base, they downloaded the Crosley and drove into town in style.

On takeoff, cadets were seated in the tail section of the B-36, an area filled with built-in cots for sleeping. Missions in the B-36 ranged from twelve to fifteen hours, so the crew could rest on the cots while airborne. After getting airborne, cadets took turns sliding down a three-foot diameter, fifty-foot-long metal tube from the back end of the aircraft, through the upper section of the bomb bay, and into the cockpit. We stood behind the pilot and copilot and looked out at the massive wings and multiple engines.

During the second week, the other cadet with whom I had driven to Fairchild was injured during physical training and was unable to finish the tour. Two cadets down, one to go.

Actually, I was not feeling very well myself by the second week. For the entire time I had been in camp, I had little time or inclination to have a bowel movement. My intestines felt extremely full, and I had a hard time closing the zipper on my pants and connecting my military belt. Another cadet offered to give me medicine for constipation, a white liquid in an unmarked bottle. I drank all the fluid gladly, but did not see any immediate results.

The highlight of the entire camp was a flight in a Lockheed T-33 T-Bird. This aircraft was a two-seated basic jet trainer used in Air Force flight schools. It was the aircraft I would later fly when I qualified for Air Force silver wings. Other cadets blustered about the acrobatic maneuvers they had asked the T-33 pilot to perform. Some talked about an aileron roll, a loop, or even a spin.

All of the maneuvers sounded fine to me. After strapping a parachute on, attaching the seat belt and shoulder harness, and putting on a jet flight helmet and oxygen mask, I became apprehensive and my body felt restricted and uncomfortable. The T-33 pilot wrote my first name in grease pencil on a rearview mirror attached to the front part of his canopy. He flew so many cadets that he needed a method to keep track of each name. As soon as we got airborne, the feeling of speed, acceleration, and maneuvering Gs confused me. I didn't understand the communication when the pilot transmitted on the radio, and nearly every part of my body hurt from the uncomfortable flight equipment. I hoped and prayed the pilot would not do all the acrobatics I had requested. To this day, I have no idea where we flew or what maneuvers we went through; the Gs pulled my head down so far that I saw only the cockpit floor.

By the third week, all the cadets in our group were aware of my highly irregular bowel habits. I was in great distress, but didn't want to be the third and last of my group to be disqualified. Because my distress had kept our group in good spirits and laughter, they elected me as our laundry and morale officer, similar to Ensign Pulver, played by Jack Lemmon in the movie *Mr. Roberts*. Later I found out that the white liquid was Kaopectate, used for diarrhea, not constipation.

I did complete the course, but my final evaluation stated that I did not take the military seriously. I appeared to be a "goof-off" and was more interested in having a good time and joking around than in Air Force training. This grade was absolutely correct; I wanted to be a pilot much more than a military officer.

After I graduated from college I would be called into the USAF sometime during the next twelve months—too many cadets graduated for all to go into the Air Force at the same time. In the spring of my senior year at college, I flew in a United Airlines Douglas DC-6 to Washington, D.C., to interview for a job as a chemist with Reynolds Metals. After landing, I took a taxi to the old Smithsonian Institution building. When I entered the main building, I saw both the Wright brothers' *Flyer* and Charles Lindbergh's *Spirit of St. Louis* hanging from the ceiling. What a sight! Every boy interested in aviation had read and reread the stories about these pilots and planes. Now I had the opportunity to observe these aircraft firsthand. The Wright brothers and Lindbergh were some of the earliest test pilots.

In a Quonset hut at the back of the main building was the rest of the aviation display. Chuck Yeager's orange X-1 *Glamorous Glennis*, the first aircraft to fly at supersonic speeds, was parked in a corner. The interior of the building was not well illuminated and looked like an auto repair garage. Still, it was a great experience to study firsthand three of the aircraft that had made a huge impact on aviation history.

Only three months remained until I would graduate from college and become a second lieutenant in the United States Air Force.

Chapter 3

USAF Flight School

The following February I married Janice "Jan" Sheehan in Omaha, Nebraska, on one of the coldest days of the year. Jan, a tall, slender young lady with auburn hair, was of Irish and German descent. She was also born and raised in Nebraska, and we had met just over a year earlier, on New Year's Eve, when I had gone back for a visit to my hometown of Grand Island, Nebraska. I was still in college and she was a senior in high school. After graduating from college, I'd taken a job as a chemist for seven months at Firestone Tire and Rubber Company in Akron, Ohio. By Christmas I was called to active duty in the Air Force and ordered to report to Lackland Air Force Base in San Antonio, Texas, at the end of February 1958. Jan and I decided to get married during the middle of February and start our new life together.

Jan would be nineteen the following month, very young in retrospect to be married, but she was now an officer's wife. We loaded my black 1948 Chevrolet business coupe with our wedding gifts and headed south. Jan referred to my Chevrolet as the "Humphrey Bogart" car since Bogart often drove a similar car in his movies. As we drove through Kansas and Oklahoma, the weather warmed and the sky cleared. We were newlyweds off to see the world.

I entered active duty in the United States Air Force as a second lieutenant. The course at Lackland was very similar to the Air Force ROTC summer camp, with more marching, more physical training (PT), and more Air Force history. With a clothing allowance from the Air Force, second lieutenants purchased their officer uniforms. These uniforms were much nicer than the uniforms from college ROTC—and they didn't have any bubblegum on them.

While we were in San Antonio Jan and I fell in love with a sports car owned by one of my classmates who was from the state of Washington. It was a robin's egg blue 1958 MGA roadster made in England and only six months old. We made the first purchase of our married life and became indebted after I had been employed for only a couple of weeks. We cruised the Alamo, the shrine of Texas liberty, on early spring Texas evenings driving with the top down. We were finally out of the cold and snow of Nebraska.

My next assignment was primary flying school at Bainbridge Air Base in Bainbridge, Georgia, a small town in the Deep South, only twenty miles north of the border with Florida.

Back in the 1950s, traveling across the United States by automobile was a different experience than it is now. The interstate highway system we now take for granted didn't exist. Even federal highways were only two-lane roads that passed through the center of every town in their paths. Franchise motels and restaurants did not exist. As we traveled, we experienced the local color of the state, eating food that was very particular to just that area.

We drove east leaving Texas, Jan driving the MGA roadster and I following in the Chevrolet business coupe, playing the part of Humphrey Bogart. Crossing the Gulf Coast states, my young bride attracted a great deal of attention as she drove our sports car, top down, her hair blowing in the breeze. Driving a car with a Washington state license plate, she appeared to many truckers to be a long way from home. Little did they know that Bogart was following close behind. Jan instantly made friends at every restaurant and gas station in Louisiana and Alabama, and before long she could amuse me by speaking in the local dialect.

Life in a small town in Georgia was vastly different than what we had experienced in Nebraska. We lived on the first floor of an old, furnished duplex without air conditioning, TV, or any radio reception. Each of the four categories of poisonous snakes found in the United States lives in Georgia. Our landlord killed several on his property near our duplex while we were there. Jan remembers the Georgia chain gangs. As many as fifteen prisoners, dressed in black-and-white–striped uniforms, were chained together as they worked on the road near our home. At each end of the chain gang, Georgia state troopers stood guard with shotguns.

A civilian company called Southern Airways contracted with the Air Force to provide flight instruction for student Air Force pilots. Carl Smith, my civilian flight instructor, had been a Boeing B-29 Superfortress pilot during World War II. He taught me to fly my first military aircraft, the Beechcraft T-34 Mentor, a plane similar to the civilian Bonanza. The T-34 had a single reciprocating engine and retractable landing gear. To me it was the most beautiful aircraft in the sky at the time. After I had only thirty hours of flying time in the T-34, an Air Force military check pilot gave me a thumbs-up on my final proficiency flight exam. The next training aircraft was the Cessna T-37 Tweety Bird, a noisy twin-engine jet aircraft also known as the "six-thousand-pound dog whistle." I had finally become a jet pilot and now wished Bob Preston, my childhood flying friend, could see me with my symbolic "whistle."

Our flight school class was known basewide for having some of the most colorful student pilots. One student, 2nd Lt. Gerald J. Nutting, was born and raised in Brooklyn, New York. Jan said she could not understand a word he said. She thought he had a speech impediment and questioned how he had qualified for flight training. His voice was fine; he just talked "Brooklynese."

While flying solo in the T-34 over our local flying area, Nutting practiced simulated forced landings. For safety reasons, Air Force rules prohibited student pilots from practicing forced landings without their instructor on board. But Nutting didn't always follow the rules. After completing one of his illegal simulated forced landings, he added engine power and started to climb. Nutting had not visually cleared the sky above and in front of him. His aircraft propeller sliced through the aileron of another T-34 that was also flown by a student. Both planes were extensively damaged, but neither pilot was hurt. It was the first close call any student in our class experienced. Nutting was given extra briefings and instruction.

Nutting's notoriety continued. One evening, after he had been drinking, he drove his brand-new Chevrolet at high speed through the display area of a local pottery shop, scattering clay ducks, geese, chickens, and other assorted animals and statues across the Georgia landscape. The owner complained to the Air Force, and Nutting was in trouble again. Several months later, Nutting was eliminated from flying school for poor flying performance, which in the long run probably saved his life.

Another student in my class, 2nd Lt. Terry Uyeyama, was Japanese by ancestry but grew up in New Jersey before joining the Air Force. On one of his solo flights in the T-37, he forgot to monitor the amount of jet fuel remaining in the aircraft. He finally landed after using 300 gallons of the 311 on board. This was an extremely low fuel state, especially for a student. After receiving hours and hours of reeducation from his instructor, Uyeyama was scheduled for another solo flight. The following day, ignoring the pleas of the fainthearted, Uyeyama successfully proved that the T-37 could and would fly on fumes, landing this time with only two gallons remaining. Other student pilots asked him whether his father had been a kamikaze pilot in World War II who had not been concerned about the amount of fuel on board. Uyeyama didn't think the comment was funny.

Uyeyama completed flying school and later flew the F-4 Phantom in combat during the Vietnam War. Ten years after being classmates in flying school, I saw him again in the Officers Club at Udorn Royal Thailand Air Force Base (RTAFB) in Thailand just after I first arrived to fly combat. Within a week, he went down on a mission over North Vietnam and spent five years as a POW before returning during Operation Homecoming in 1973. I always wondered whether he had really been shot down or had just run out of fuel.

Besides flight instruction every day, academics and physical training filled up our student pilot time. With my math and science background, the lessons on aircraft aerodynamics, navigation, and weather were easy and I received high grades.

On the other hand, the physical fitness training under the hot Georgia sun and high humidity was a challenge. We ran miles and miles around the track, did push-ups until we were exhausted, and then climbed ropes every day in temperatures in the mid-90s. It was not much fun, but I was in the best physical condition of my life.

Toward the end of our six-month flying school, the three classes of students were scheduled to compete against each other in a track meet. Our class had won the softball tournament and had set a goal of winning the track meet as well. Many of the student pilots had been college athletes and played sports at a college level of performance. Now, with all the physical training, they were in top-notch shape. Each class was allowed only two entrants in each event, meaning that six student pilots would be competing in every event.

Since my days as a youngster, running had always been my sport. However, the best I had placed in our class practice competition was third in the 200-yard dash. With all of these student pilots in such good physical shape, our student class leader thought he needed to use me only on the four-person mile relay team.

By the day of the track meet one of my classmates who was scheduled to compete in the 200-yard dash had been eliminated from the flying program, and I was selected to replace him. The 200-yard dash was conducted on a standard oval track. The event has a staggered start on the curve and finishes on the straightaway. I was really pumped to have the opportunity to help our class win the meet. When the starting gun fired, I shot out of the chocks, afraid I would be the last to cross the finish line. Due to the staggered start, I didn't know my placement until I entered the straightaway. To my surprise, I was leading the race and crossed the finish line in first place. Our class was now tied for first.

The 100-yard dash was the next event, conducted entirely on the straightaway with no turns. Officials were registering the competitors and making sure the track was clear. Standing at the finish line, I could sense the excitement building. Our student class leader was determined that we win the track meet. Winning would place us in good light with the military officers, and the check pilots might make the final T-37 check flight easier for us.

The official at the start of the 100-yard dash announced that my class had only one entrant. For unknown reasons, our second entrant was missing. The official said he could start the race with only five entrants but would hold the start for a few seconds to let our class find a substitute. Our student leader asked me to run. Fortunately, I still had my tennis shoes and shorts on from the 200-yard dash. Although I had no time to warm up or even practice a start, in the Georgia sun you never really cooled down. Arriving at the starting line, I was already out of breath from the previous race. With the comfortable thought of the 200-yard dash win fresh in my mind, however, my feet didn't touch the ground after the gun was fired. I had one of those peak experiences that I was to read about later. I seemed to fly over the track using little effort or strength and glided across the finish line far ahead of my competitors. Our class won the track meet.

All of the student pilots in my class were winners, ready to graduate from primary flying school and move on to basic flying school. Jan and I traded our 1948 Chevrolet business coupe for the last month's rent in the duplex. We loaded our belongings into the small MGA roadster and headed back west to Texas. On this trip we were both in the MGA. I didn't see any truckers follow us this time.

My new assignment was basic flying school at Webb Air Force Base in Big Spring, Texas. Jan and I arrived only a day before my class started, and we checked in to the Big Spring Inn, a motel just off the end of the Webb runway.

Big Spring in west Texas was flat and dry, oil country. You could smell the sulfur-like fumes from the Cosden Refinery across town, even from our motel. Jan and I told the motel manager that it smelled awful. We were corrected; he said it smelled like money. We got the point.

Texas had been in a drought for some time; often a strong dry wind from the north blew red soil from Oklahoma causing what they called "dusters." The sky turned light brown, the motel windows rattled, and dust covered everything.

For six months I learned to fly the T-33 T-Bird, a single-engine jet trainer developed from the P-80 Shooting Star of Korean War fame. On successful completion of this flying training, I would be awarded silver pilot wings and be fully qualified as an Air Force military pilot. The flying instruction was really accelerated at Webb compared to primary training back at Bainbridge Air Base. Besides learning to fly takeoff and landing patterns and acrobatics as we did in the T-37, in the T-33 we also flew at night, on instruments, in formation, and practiced cross-country navigation.

Before we realized it, Jan and I had been living in the motel for a month. Since we had no furniture and furnished apartments were impossible to find, we decided to stay in the Big Spring Inn. Jan bought a five by seven foot rug for our living room. The room was so small we called it wall-to-wall carpeting. To finish our decorating, we purchased a very small bookcase. The motel would have to be home for the time being.

In the evenings we ate in the motel restaurant, played shuffleboard, and drank beer. Jan had not turned twenty yet, but the bartender never asked her for an identification card. After all, she was an officer's wife.

My flight instructor for the entire six months of training, 1st Lt. Lenard W. Kresheck, was just a year older than the three of us who were his students. Unlike Carl Smith, my primary instructor who seemed old in his early forties, Lenard was young and had a good sense of humor. We instantly became good friends and I felt at ease when I received flight instruction from him. Due to Lenard's easy demeanor, I quickly learned to fly the T-33 and became one of the first students in our class to solo.

Soon all three of Kresheck's students successfully completed acrobatics, night flying, and cross-country navigation and were ready to start one of the most difficult phases of flight training: instruments. With the instructor sitting in the front seat, the student pilot occupied the rear seat and learned to fly the aircraft with reference to only an instrument panel full of many aircraft and engine gauges. A piece of cloth canvas supported by bungee cords, called the hood, covered the entire inside of the canopy preventing the student pilot from seeing any of the outside world. We called this "flying under the bag."

I found basic instrument flying to be easy, but learning the complicated instrument approach procedures and instrument flight rules (IFR) was much harder. Gradually I became proficient in flying precision radar approaches called ground control approach (GCA) and fully automated approaches called instrument landing system (ILS). The ILS is an approach still used today. These precision approaches allowed a pilot to fly to a runway where the weather was as low as 200-feet overcast with one-half-mile forward visibility. We also practiced nonprecision approaches, called visual omni range (VOR) and automatic direction finding (ADF), that were used for weather conditions down to 400-feet overcast with one-mile visibility. Since it was winter, many of the practice instrument approaches were accomplished in actual weather conditions.

On my final instrument check ride given by our military flight commander, I received one of the best grades in the class: excellent. Kresheck informed me that he thought I could fly instruments even better than he could, quite a compliment from an instructor.

While I could fly good instruments, another phase of instruction proved to be more difficult. This type of flying was called formation. In this phase of flight training, a student pilot learned to fly his aircraft only a few feet away from another aircraft and to duplicate the exact aerial maneuvers of the lead aircraft. All of our training to that point had been to avoid other aircraft. It was strange to see another plane so close, and the natural reaction was to pull away. I had flown several flights with Kresheck in the rear cockpit guiding me gently on the small flight control and throttle movements required to stay exactly in position. It was very easy to overcontrol the aircraft with too large an input and end up looking like I was on a roller coaster.

I had not soloed yet but was scheduled to fly formation with 2nd Lt. Dave Mosby, one of Kresheck's other students. Mosby's flying performance had been superior up to that point; he had even been cleared to fly solo in formation flight.

Kresheck was ill so 1st Lt. Albert E. Haydel Jr., another flight instructor, was assigned to fly with me. The plan was for Dave Mosby to fly solo formation on my wing for the first half of the flight and for me to fly supervised formation on him the last half with Lieutenant Haydel sitting in my backseat. Haydel was a very accomplished, but difficult, instructor. He had an air of superiority and treated all students harshly. On practically every flight that he flew with a student who was not his full-time student, he gave them a failed grade. He usually recommended that the student be eliminated from flight training. With that recommendation, a student would be scheduled for an evaluation flight with the flight commander. Failure on this flight and a student was gone, a quick process.

After preflighting my aircraft, Haydel and I strapped into our parachutes, attached our seat belts, and donned our flight helmets. External electric power was applied to the aircraft and the ultra high frequency (UHF) radio turned on to listen for Mosby to call us and report that he was ready for engine start. Our aircraft was equipped with a hot-mike intercom allowing Haydel and me to communicate with each other just like on a telephone. We waited for several minutes for Mosby to contact us.

I was very nervous about the upcoming flight. In fact, my knees were shaking just thinking about it. I did not feel I could perform well while receiving the verbal abuse and swearing Haydel was known to exhibit during flight. This was sure to be an unsatisfactory mission, and I expected that soon I would be up for a dreaded elimination flight with the flight commander.

We continued to wait for Mosby to call us on the radio. Other instructors and students were talking on the radio, but there was no communication from Mosby. Suddenly I saw Mosby running toward our aircraft. He jumped up on the right wing and leaned into the back cockpit to talk to Haydel. Because of our hot-mike intercom, I could hear both sides of their conversation.

During preflight, the pilot visually checked that each of the aircraft's seven fuel tanks was full of fuel and that the caps were securely locked. Evidently,

Mosby had accidentally dropped his small metal screwdriver in the fuselage fuel tank. The screwdriver sunk to the bottom of the ninety-five-gallon tank, about four feet below the cap. The aircraft was now out of commission and grounded by maintenance personnel. Haydel was livid with anger.

"How could you do such a dumb thing?" he demanded of Mosby. "How could a student be so stupid?"

Try as he might, Mosby could not explain how he had dropped the screwdriver into the fuel tank. Haydel became more and more angry and spoke louder and louder. I listened to the conversation, one-sided as it was, not saying a word.

Haydel ordered Mosby to contact the maintenance department and request a spare aircraft. Mosby ran off, and Haydel continued to rant and rave on our intercom. He was swearing and called Mosby every degrading name he had ever used on a student. On the other hand, I was very, very quiet.

Our block of flying time was slowly being used up as we waited for Mosby to check in on the radio. Finally, he called and reported that he was ready for engine start.

I taxied to the runway for takeoff; Haydel continued to give nonstop instruction and corrections to Mosby both on the UHF radio, which he could hear, and the intercom, which he could not. Haydel said nothing about my performance.

During the entire flight, Mosby was criticized and critiqued by Haydel. Mosby's formation position looked correct with smooth control action from my position and point of view. As a matter of fact, I thought he looked great. When Mosby took over the lead, I flew his wing through all of the maneuvers with Haydel making absolutely no comment on my performance. He could not shake the screwdriver incident from his mind.

I landed the plane, taxied to the parking area, and shut down the engine. Haydel departed our T-33 without saying a word and walked toward Mosby's aircraft.

The next day I read my flight evaluation report. I was graded average in all categories. Kresheck was amazed.

"You must have flown very well to get an average grade from Haydel," he remarked. "I'll put you up for solo formation on your next flight."

From that flight on I relaxed while flying, the key to good formation flying. I received another "excellent" from the flight commander on the final formation check flight.

I never had a chance to thank Mosby for taking the heat off of me. Dropping the screwdriver in the fuel tank was no doubt an extremely embarrassing incident for him. But he still ended up graduating fourteenth in our class of sixty-five, one in front of me. At the end of our training, the Air Force selected him to fly the North American F-100 Super Sabre in Tactical Air Command (TAC), and I'm sure he had a brilliant flying career. Haydel, on the other hand, perished in an F-100 several years later.

Lenard Kresheck stood next to me when Jan pinned the Air Force silver wings on my uniform. He recommended to the pilot selection board that my name be added to a list as a jet fighter pilot. He was a vital part of my dream coming true and another step in chasing the whistle.

Jan and I packed our belongings again, which was getting to be routine now as this was our third move in twelve months. The military-supplied shippers sent our small rug and bookcase on to Georgia. My next assignment would be advanced flying school at Moody Air Force Base in Valdosta, Georgia, flying the North American Aviation F-86L Sabrejet.

Our MGA roadster made another trip east out of Texas crossing the Gulf Coast states. Valdosta, Georgia, was about eighty miles east of my primary flying school in Bainbridge, where I had flown six months earlier. Our Moody local flight training area encompassed the Okefenokee Swamp, a huge wildlife refuge inhabited mostly by moonshiners and alligators.

The F-86L was an advanced version of the F-86A, a plane that had been very successful in shooting down MiGs in the Korean War. Modified with radar, air-to-air missiles, and an afterburner, the F-86L was a defensive fighter stationed at Air Force bases along both East and West coasts and the northern border with Canada. Their primary duty was to intercept and destroy enemy Russian bombers were they to attack the continental United States.

Since it was a single-seated aircraft, I learned to fly the aircraft and operate the radar at the same time. This required high skill on the part of a pilot and good instrument proficiency. As a matter of fact, 1st Lt. Cloyce G. Mindel, one of my fellow students, got disoriented on a night practice intercept mission, lost control of the F-86L, and crashed into the Okefenokee. This loss of a pilot friend was the first of many I would experience in my forty-nine-year flying career.

Just as we had done the year before, Jan and I spent a long hot summer in Georgia. Our rented house did not have air conditioning so we bought a large window fan that we moved from room to room with us. At night we tossed a coin to see who got to sleep directly in front of the fan. The fan rattled and shook, keeping us awake at night. Sometimes we got out of bed in the middle of night and drove the MGA roadster back and forth through low points in the highway where the air was just a few degrees cooler. In August Jan became pregnant. You could blame it on either hot Georgia nights or a noisy fan.

One entire training flight in the F-86L was devoted to supersonic flight. Using afterburner, I climbed the Sabrejet to 40,000 feet, leveled off, and accelerated to a speed of 0.9 Mach. Still in full power, I rolled the aircraft inverted and pulled the nose straight toward the ground, aiming at imaginary moonshiners in the heart of the Okefenokee Swamp. At about 30,000 feet, the F-86L exceeded the speed of sound and created a sonic boom for all the alligators to hear. For me, there was no physical sensation in the cockpit when I broke the sound barrier; all I could see was a jump in the Mach meter. In seconds it was over, the boom ended, most of the fuel was depleted, and I returned to the Moody traffic pattern. For breaking the sound barrier, each pilot received a North American Aviation Mach Buster card and tie tack. I had climbed another rung on the ladder to becoming a fighter pilot and was in pursuit of an even bigger whistle.

With eighteen months of active duty in the Air Force, I was promoted to first lieutenant. Now that we had a little extra cash, on weekends Jan and I drove all over southern Georgia and northern Florida. The English-built MGA roadster still attracted a lot of attention in small southern towns in the late 1950s.

Once, a fan belt broke while we were on the road. A Georgia old-timer looked at it and said, "I think a belt off my old washing machine would work." It did. We enjoyed southern hospitality, just not the weather.

In October, as the end of my advanced flying school program approached, I anxiously awaited my first permanent military base assignment. The new assignments were based entirely on class standing, not on recommendations from instructors. All thirty-nine graduating fighter pilots gathered in the base auditorium. Thirty-nine Air Defense Command squadrons were listed in chalk on a blackboard, with appropriate location and aircraft type shown. The top-ranking pilot had his choice of any squadron listed. After he chose, the squadron he selected would be erased and the next highest-ranking pilot got his choice.

My position was number nine, high enough to ensure I would get a prime fighter assignment. But I had raised my personal expectations, chasing the whistle again. Now I wanted to fly the Century Series category of aircraft some place other than the South. Century Series aircraft were the newest versions of Air Force fighters that could fly faster than the speed of sound in level flight, not just in a dive as the F-86L did. Century Series jets were the F-100, F-101, F-102, F-104, F-105, and F-106 aircraft.

Luck came my way. The MGA roadster would cross the entire United States, this time to California. My new assignment was to the 84th Fighter Interceptor Squadron (FIS) located at Hamilton Air Force Base just thirty miles north of San Francisco's Golden Gate Bridge.

Chapter 4

Defending the Golden Gate

Jan and I drove our 1958 MGA roadster from Georgia to California in December 1959. She was now five months pregnant and looking forward to getting settled in a home and to the birth of our first child. As we came down the western side of the Sierra Nevada range into the Sacramento Valley, we saw the winter greenery of the West Coast. What a pleasant change from the snow of Nebraska and the humidity of the South. Within a couple days we had rented a partially furnished two-bedroom house in Novato, a town a few miles north of the base. The house was nothing spectacular, but we were now living in sunny California and I was going to be a fighter pilot.

I was eager to visit Hamilton Air Force Base to see firsthand the flight line, hangars, and supersonic aircraft. Hamilton was the home of the 78th Fighter Wing, which had achieved a distinguished combat record flying P-51 Mustang missions out of England over Germany during World War II.

Two fighter squadrons were located on the base. One was the 83rd Fighter Interceptor Squadron (FIS) equipped with the Lockheed F-104A Starfighter. During the previous year, pilots in the 83rd FIS had set a world speed and altitude record with the aircraft at Edwards Air Force Base. The F-104A was a red-hot Mach 2 interceptor armed with two Sidewinder heat-seeking missiles.

My squadron was the 84th FIS, which had recently given away their antique Northrop F-89J Scorpions, an ugly subsonic plane nicknamed "the gravel-gobbler" due to its drooped nose and low air intakes, to start flying the McDonnell F-101B Voodoo. The F-101B was a twin-engine two-seated 45,000-pound jet interceptor armed with two nuclear rockets and two heat-seeking missiles. The plane was about twice as large and fast as the F-86L Sabrejet I had recently flown in advanced flying school.

It was now the middle of winter and coastal fog covered the base almost every day. New pilots, recently assigned to the combat squadrons, were unable to check out in the interceptors until the weather improved. Six new guys, including myself, were assigned to the T-33 target group, otherwise known as the holding tank. We were all waiting for a chance to fly Century Series aircraft. This target group flew missions as simulated enemy aircraft making surprise attacks on the base.

It was in this group that I met Capt. William A. "Bill" Anders, twenty-seven, from San Diego, another pilot assigned to the 84th FIS. Anders, a 1955 graduate of the U.S. Naval Academy at Annapolis, had most recently flown the F-89 in Iceland. He had been waiting a couple of months for his turn to check out in the F-101B.

Three pilots assigned to the 83rd FIS were waiting to fly the F-104A: 1st Lt. Jim Milner, a fellow AFROTC summer cadet whom I had met at Fairchild Air Force Base, Washington, in 1956; 1st Lt. John Taylor; and 1st Lt. Ken Mason. All were excellent pilots and we enjoyed flying the T-33 together. But we were also anxious to get checked out in the supersonic interceptors and become combat ready.

Bill Anders and I got to know each other quite well. He and his wife Valerie lived in base housing and had several children. Anders owned an old Volkswagen bug and enjoyed the challenge of testing its crosswind-handling characteristics driving across the Golden Gate Bridge when the wind was blowing up a gale. It was all he could do to keep the VW on the road as wind gusts pushed it around. Little did I realize then how he would keep his space capsule on the road a few years later flying a historic flight to the moon.

The Winter Olympics were held at Lake Tahoe in early February 1960. Anders thought this was a once in a lifetime opportunity and that we just had to attend. Winter sports were new to Jan and me, but the trip did sound attractive. All four of us squeezed into Bill's bug for the four-hour ride to Tahoe. Bill and Valerie were in the front seat; Jan, who was now seven months' pregnant, and I were in the backseat. How Jan managed to keep a smile on her face and a sense of humor I will never know, but we had a wonderful time.

Soon Anders started to check out in the F-101B as did Ken Mason in the F-104A. Mason, tall and athletic, was an excellent basketball player and a nondrinker who did not party, the custom for fighter pilots of that day. On Mason's second checkout flight, his first mission in the front seat, the General Electric J79 jet engine failed while he was in the landing pattern. The aircraft was equipped with downward ejection seats and he and his instructor pilot ejected over San Pablo Bay. Mason's parachute opened and he landed in the bay. The water was very shallow where he landed, only a couple of feet deep. However, the wind was blowing strongly and the parachute canopy pulled Mason under the water. Even though he was in excellent physical condition, he couldn't prevent himself from being dragged for some distance and he eventually drowned. Flying Century Series aircraft was very dangerous.

Standing on the flight line parking ramp shimmering in the early morning misty fog, the F-101B Voodoo exuded sleekness and force. It appeared menacing, but it was also streamlined and very aerodynamic. An old Air Force adage stated that "the nicer an airplane looks, the better it flies."

The Voodoo looked nice, but how did it fly? Before long I got a chance to find out. I didn't realize how big the F-101B was until I climbed the ten-foot entrance ladder and looked back down at the parking ramp. The cockpits of my jet training aircraft were only a few feet off the ground—now I felt like I was already airborne. After helping me attach my parachute and shoulder harness, the crew chief climbed back down the ladder. As he removed the ladder and I closed and locked the canopy, a feeling of isolation came over me. I was on my own now—and I was going to fly this huge beast by myself. I did have a radar intercept officer (RIO) in the backseat, but he was not a pilot and didn't know how to fly.

For its time the Voodoo was one of the highest-performing fighter warplanes in the Air Force inventory. As a first lieutenant, just six months out of flight school, I found the aircraft a lot to handle. With two big Pratt & Whitney J57 turbojets plus afterburners pumping out about 25,000 pounds of thrust, it was especially important that I stay mentally ahead of the aircraft. My flight instructor had warned me to "never let an airplane take you somewhere your brain didn't get to two minutes earlier." It would take a lot of training for me to keep up with the F-101B.

While the Voodoo was certainly a fighter from a speed and acceleration perspective, its very high wing loading (110 pounds per square foot of wing surface) meant that it handled like a bomber when flying subsonic at lower airspeeds with a full load of fuel. It required a light touch and good hand-eye coordination. No daydreaming in this machine or it would certainly get away from you.

Both engines were easy to start, and little power was required to taxi. Once I lined up with the runway for takeoff, I checked full power on each engine separately—leaving the other in idle. The brakes held the aircraft stationary with one engine at full power, but both engines at full power produced so much power that even though the brakes would hold, the tires would rotate on their rims.

After the engines, hydraulics, and flight controls were confirmed ready for takeoff, the Voodoo was ready to roll. I advanced both throttles to 85 percent rpm, depressed the nose-wheel steering button on the flight-control stick, and released the brakes. As soon as the F-101B started to move, afterburner was selected separately on each engine. Even with my jet helmet on I heard a loud "boom-boom." Acceleration down the runway was fantastic.

I was airborne in eight to ten seconds at a speed of 175 knots (200 mph). This was as close to a Navy catapult launch as you could find in the Air Force.

Civilian light aircraft have direct wire cables from the control surfaces to the flight stick in the cockpit. As the aircraft increases airspeed, air pressure builds up and the controls become stiffer. Because of the Voodoo's weight and high speed, the F-101B's flight controls required three thousand psi (pounds per square inch) hydraulic pressure to move against high air loads. The pilot flew the aircraft using an artificial feel system, just like power steering in a car.

The Voodoo's flight controls were extremely light, especially during takeoff. A pilot had to be very careful not to overrotate or point the nose too high at liftoff and stall the plane. The first such accident in the aircraft occurred in 1956 when Maj. Lonnie R. Moore, a Korean War jet ace with ten MiG kills, took off from Eglin Air Force Base, Florida. He overrotated, crashed, and was killed. The death of this highly decorated Korean War jet ace made headlines and alerted the Pentagon brass that even experienced pilots could have difficulty with an aircraft that pilots were now calling "that great big, wonderful bastard of an airplane."

The problems weren't over when I got airborne either. The aircraft accelerated so fast that I could easily exceed the maximum gear-down speed of 250 knots. If I didn't get the landing gear handle up immediately when airborne, air pressure would hold the forward retracting nose gear down and the main gear doors could even rip off. Taking off in the Voodoo was like running a 100-meter dash—your heart pounded and you were almost out of breath when it was over.

Even after takeoff the acceleration of the F-101B was impressive. Using a 30-degree climb angle, I accelerated to 0.85 Mach and climbed to 35,000 feet in just three minutes. In only a couple of minutes more of afterburner power, I accelerated to 1.73 Mach, the maximum speed for the Voodoo. Frequent training missions were absolutely mandatory for a pilot to ever feel proficient and comfortable handling this aircraft. Some pilots in training status never physically adjusted to the fast-paced motions of the F-101B and were eliminated from flying the plane. They were then trained to fly slower aircraft.

On the other hand, no pilot would ever relax and adjust to one notorious feature of the Voodoo. This feature practically prevented the aircraft from ever becoming a combat-ready aircraft in the Air Force inventory. If you want to cause the hair on the back of an F-101B pilot's head to stand up, whisper "pitch-up" in his ear. The T-tail configuration of the Voodoo was the cause of pitch-up. At high angles of attack, the wing disrupted and blocked the airflow over the horizontal tail, causing it to lose lift. This resulted in a severe nose pitch-up, sometimes followed by a potentially disastrous spin. High angle of attack can result from flying too slow in straight-and-level flight—hence the high takeoff and landing speeds. It was critical for the pilot to fly the flight manual recommended speeds. If he was too slow a pilot would pitch-up in the traffic pattern, which was sometimes a fatal mistake if his ejection seat or parachute failed.

Excessive angle of attack could also be experienced during high-G maneuvering, even at high speeds. To prevent pitch-up and a resulting high accident rate, McDonnell Aircraft Corporation engineers modified the flight controls. As a pilot neared the pitch-up boundary in a high-G turn, a steady tone or horn sounded in his jet helmet headset warning him that he was nearing a dangerous flight condition. If he continued pulling back on the flight-control stick, causing an even higher angle of attack, a pusher would physically force the control stick forward, preventing pitch-up. Unfortunately, this required a pilot to keep one eye on the G meter and one eye out the windscreen to get maximum turn in the Voodoo. In hard, high-rate maneuvering turns where a pilot was looking over his shoulder to prevent an opposing fighter from flying in his six o'clock or tail position, the F-101B did not score well.

One feature of the Voodoo always made me feel uneasy as I cruised at high altitude. The F-101B clamshell canopy was hinged at the rear and enclosed both the pilot and the backseat RIO. Due to the canopy's size and weight, aircraft electrical power was used to open, close, and lock it. Three separate electrical switches (one external and one each for the pilot and RIO) were interconnected so that the canopy could be controlled from any of these positions. The aircraft design engineers suspected this arrangement was a safety hazard: either the pilot or RIO might lower the canopy without observing that the other crew member still had an arm or hand resting on the canopy sill, so an override feature was installed. This feature resulted in any command to open the canopy overriding any command to close the canopy, thereby preventing serious injury to a crew member. But electrical circuits are known to short out, creating the possibility that the canopy would erroneously get an open signal. It would then start to extend in flight, causing it to rip off and explosively decompress the crew.

One pilot was flying an F-101B night practice intercept mission over the Pacific Ocean when the canopy mysteriously blew off. Due to extreme noise and windblast, the RIO could not communicate with the pilot over the intercom, and he could not raise his head high enough to visually confirm that the pilot was still in his ejection seat. The RIO assumed the pilot had ejected because of some unknown aircraft problem, and so he did likewise. The pilot, still in his seat, slowed the speed of the aircraft and attempted to make contact with the RIO without success. It was only after landing that the pilot discovered the backseat empty. No trace of the RIO was ever found.

After my first night flight in the F-101B, I came home late one evening to eat dinner and tell Jan about my flying experiences.

As I started to relax, she responded by saying, "Tonight's the night."

"What do you mean?" I asked.

"I'm in labor, and we need to go to the hospital," she explained.

Randall Alan "Randy" Marrett was born several hours later in the base hospital. After five days in the hospital, we brought him home in our MGA roadster. The sports car was far too small for a family of three. Although my days driving a roadster were about to be over, I was now flying a sports car of a plane.

Being stationed at Hamilton, just north of San Francisco, was considered choice duty for a young first lieutenant pilot. The weather was very mild compared to North American Air Defense Command (NORAD) bases in the northern states, and San Francisco offered a great variety in the way of leisure and entertainment. If a person had to go to war, defending the Golden Gate Bridge was as nice as you could find.

Before long I was fully qualified as a combat-ready pilot in the F-101B and standing alert with my flight. There were four flights in our squadron, each of which consisted of six pilots and six RIOs. Three times a month our flight was assigned a three-day alert duty that required us to be ready to fly four aircraft loaded with nuclear weapons. We were on alert straight through the entire seventy-two hours, sleeping and eating in the alert hangar that held the aircraft. Higher Air Force headquarters could give orders to fly, or be scrambled, at any time, and we were required to be able to be airborne in one hour. Two additional Voodoos were each loaded with only two heat-seeking missiles. Required to be airborne in five minutes, they were called the "Hot-5" aircraft.

I felt comfortable handling the Voodoo, had learned our tactics, and was even leading flights myself. On one occasion I was too comfortable. I was scheduled for a practice intercept mission with a takeoff at dusk. I called Jan and told her to step outside and look into the sky when I flew over; I had a surprise for her. Our house was practically under Voodoo's flight path. The sun had just set and I blasted off with afterburners in full flame. After getting airborne, I retracted the landing gear and accelerated to 350 knots flying speed. I then rotated the aircraft nose up to 30 degrees above the horizon and abeam my house, and, in full view of Jan, executed a precise 360 degree aileron roll. Although the maneuver was not dangerous, it was illegal under Air Force regulations. I was well clear of the base and sure the Hamilton control tower operator did not see me. So far, so good, I thought to myself. I flew the mission, landed, debriefed the flight, and went home. Jan was duly impressed.

The next morning when I reported for duty, I was informed the squadron commander wanted to talk to me. A deputy county sheriff had observed my fantastic display of aerial acrobatics and phoned the base to report it. He gave military officials the exact time and place of the occurrence and I was identified as the offender. I was in deep trouble.

The Air Force encouraged aggressive behavior in young fighter pilots, but it still needed to keep control over them. I was grounded for a week and assigned an additional duty as the assistant flying safety officer. I was required to read accident reports, conduct inspections, and write up unsafe conditions in our squadron. I found the job to be a great educational experience and a chance to learn another side of aviation.

Before long, an opening occurred for attendance at a weeklong Jet Engine Accident Investigation course at Sheppard Air Force Base in Texas, and I was selected to go. What started out as a punishment became a benefit because I learned about the inner mechanical working of aircraft. This education later helped me as a test pilot: I was the official Air Force accident investigator of an F-101B crash in Oregon, a T-33 crash in northern California, and an NF-104A Starfighter accident at Edwards Air Force Base.

After attending the course I continued to be on Hot-5 and nuclear alert. One of the roles of the Hot-5 crew was to check out unidentified aircraft. Commercial jet airliners coming from Hawaii and Alaska returned to the continental United States following published air lanes over the Pacific Ocean. In the early 1960s, airlines had just started jet operations and their pilots had not flown many flights with their new aircraft over those two thousand miles of ocean. Occasionally these aircraft would mistakenly fly off course far enough that they were declared unknown by USAF Air Defense Command. F-101Bs from Hamilton were scrambled to intercept the airliners and confirm by visual identification that the unknowns were friendly.

Occasionally, the aircraft on Hot-5 would be scrambled at night and vectored by our ground radar controllers toward the unidentified aircraft. Since the two Voodoos were each armed with heat-seeking missiles, if the unknown aircraft turned out to be enemy, it could be shot down. Although our weapons switches were in the safety position, they could be changed in only a few seconds to arm and fire the missiles.

I was scrambled on several occasions to check out such aircraft. In such events, the RIO locked his radar onto a blip on the radar display. As pilot I flew the F-101B and followed a computer-generated steering dot on my display, which would guide us into a tail chase. The RIO reported over the intercom how fast I was closing on the unknown aircraft. Then I gradually slowed the aircraft to synchronize speed about 200 feet behind and slightly below the right side of the unknown aircraft. Normally, we observed the airliner's external flashing lights several miles out. They could not see us, however, because our external lights were off. Gradually we closed to about 50 feet off the airliner's right wing. By then we saw passengers through the cabin windows, but they and their flight crew had no knowledge we were just a few feet away.

The RIO called out, "Ready for ID light."

On my command the RIO switched on an eight-inch sealed-beam spotlight located on our left fuselage near his cockpit. When the light illuminated the airliner's tail, we recorded the identity of the carrier and its registration number, which we radioed to ground controllers. Meanwhile, the window shades began to open on the airliner and frightened passengers would gawk at the bright light stalking them in the pitch-black night over the ocean. We were probably the cause of many unidentified flying object (UFO) sightings in the 1960s!

Soon I got to fly the F-101B on what was called a cross-country flight. All of our squadron intercept-training missions were flown over the Pacific Ocean immediately north and west of our base, the most likely area for a Soviet attack. The Air Force believed that pilots should understand how to navigate the high-altitude air routes across the United States and conduct landings at other Air Force bases. Our leaders encouraged these flights, since navigation could be learned on a cross-country. Wild activities were rumored to have happened on such flights, since they took place away from the direct supervision of our leaders. I looked forward to going on a cross-country flight.

Finally my flight was scheduled for a weekend cross-country on which I had the opportunity to fly as a wingman. Our plan was to make an initial flight to George Air Force Base near Victorville, California, on a Friday afternoon. There we would get to inspect F-100 Super Sabres and visit the Officers Club for happy hour. On Saturday we would tour several military bases in Southern California and return to Hamilton by Sunday evening. A great adventure was coming up. Off we went in a four-ship formation, with each aircraft carrying two heat-seeking missiles. The squadron considered our flight to be on a reduced four-hour alert status.

I experienced my first emergency in the F-101B as I came in to land at George. I extended the landing gear and the right main gear showed unsafe. Each gear (two mains and a nose) had a green light that illuminated when the gear was safely down and locked. The right main gear green light was not lit.

The bulb still illuminated when I pressed the entire light assembly in, a test function that confirmed the lightbulb was working satisfactorily. The cockpit symptoms indicated that I had a problem with either the right landing gear or its indicating system. The emergency checklist required me to use pressure from a backup nitrogen gas bottle to blow the gear down.

After blowing the gear down, I still had an unsafe indication. However, the other pilots flying in formation with me reported that my gear appeared to be safely down. I made a very smooth landing, called a "greaser" in those days. I had handled the emergency well.

"Good job," my flight leader commented.

My aircraft was now out of commission, however, and the Air Force mechanics stationed at George Air Force Base did not have experience repairing the F-101B. They had been trained only on the F-100. So the Air Force maintenance officer recommended that I fly to Oxnard Air Force Base near Camarillo, California, where another F-101B squadron was located. The flight would have to be accomplished with all three gear down and safety pins installed and at a slow speed so as not to damage the gear. Since the distance from George to Oxnard was less than a hundred miles, I would still have plenty of fuel even with the extra drag from the extended landing gear.

On Saturday morning I left my flight mates to their upcoming adventure surveying the wondrous sights of Southern California and flew alone to Oxnard with my landing gear down. Oxnard was soon in sight and I landed without difficulty. Cross-country flying had not been as enjoyable as I had hoped.

The Oxnard F-101B maintenance troops were displeased to see a first lieutenant arrive on Saturday morning with a landing gear problem, as it was their day off. A landing gear malfunction could be a long and difficult repair. First, my live missiles had to be removed from the aircraft so it could be lifted on jacks and the gear inspected. To get the missiles off, armament personnel were called. They were not very happy to see me either. I wasn't making any friends at that base and I made a mental note to never return.

The missiles came off without a hitch, were stored in a metal container, and trucked to the ammo storage area for safekeeping. It took the rest of the day to repair the landing gear, and the aircraft was returned to an in-commission status. I spent a quiet Saturday night at Oxnard with my RIO.

The Oxnard armament officer informed me the next morning that his armament crews didn't work on transient aircraft on Sunday and that I must wait to get the missiles reinstalled first thing Monday. What a disappointment: two and a half days on a cross-country and I'd had one landing gear emergency, one gear-down flight, and many hours standing around the maintenance hangar with enlisted men snarling at me—and I wouldn't get my aircraft back to home base on time. And this cross-country flight was supposed to have been enjoyable.

Early Monday morning I went back to the hangar. Armament personnel arrived with my missiles and installed them on the launcher rails. Things were looking better; at least I could finally get back home. But now there was a new predicament—the launcher rails would not retract. Since I did not want to spend any more time at Oxnard, I told the second lieutenant armament officer that I would fly the aircraft home with the missiles extended on the launcher rails. He was one of the few Air Force officers I outranked. I'd already flown the aircraft with the gear extended and it shouldn't be a problem to fly with missiles attached.

The armament officer explained to me that Air Force regulations prohibited aircraft from being flown with extended launcher rails and live missiles attached. He offered to remove the missiles.

"Okay, remove the damn missiles," I said. "I'm out of here as fast as I can get airborne."

At last I was back in the air and on my way home. I wondered whether the other pilots had had at least ten hours of flying time, made numerous strange field landings, and had a ball at the Officers Club. I remembered thinking that I'd record the serial number of this particular aircraft and make sure I never flew it again.

Soon Hamilton Air Force Base was in sight. I made a perfect jet traffic pattern, landed the aircraft exactly on runway centerline, smoothly applied the brakes, and turned the plane off the runway. That should impress the squadron pilots who were waiting and watching for me to return. I had safely handled an emergency and taken good care of the aircraft. The commanding officer and my flight commander would probably commend me.

No sooner had I shut down the engine and climbed out of the aircraft than the crew chief informed me the commanding officer wanted to see me in his office. After the aileron roll on takeoff incident months earlier, it wouldn't hurt to get positive recognition for handling the difficult situation I had been faced with for the last four days.

Both the operations officer and commanding officer were in the office when I walked in. The operations officer called for me to stand at attention in a military brace and fired off a series of questions. "Where are your two missiles?" he asked.

"Where did you fire them?" said the commanding officer.

"Why didn't you contact the command post for the past three days?" he added.

This was not looking like an awards ceremony!

"Sir, the missiles are at Oxnard Air Force Base," I humbly replied.

The commanding officer asked to see my copy of a hand receipt used to transfer government property from one organization to another. I didn't have one.

"Sir, could you contact the armament officer at Oxnard?" I offered in a trembling voice.

The commanding officer picked up his phone and called the Oxnard armament officer. I recalled some of the harsh words I had used on him only hours before. I had not been told to keep in contact with the command post, but then again I didn't ask anyone either.

The Oxnard armament officer confirmed that he had the two missiles, getting me off the hook. He would never know how easy it would have been to put me away in Leavenworth Federal Prison. Eventually the proper paperwork was prepared and I was allowed to come to rest at a military at ease. Little had I realized how easy it was to get into trouble in the military service, and how difficult it was to get out of trouble. I always treated armament officers with great courtesy and respect after that incident.

Chapter 5

Cold War

During the early 1960s there was great concern that the Soviet Union would launch jet bombers loaded with nuclear weapons and bomb cities in the United States. The Cold War was at its peak and people took the threat of a Soviet first strike against the United States with nuclear equipped intercontinental ballistic missiles (ICBMs) and bombers with great seriousness.

Each one of our F-101B Voodoo aircraft could carry two of the Douglas Aircraft Company's MB-1 Genie 1.7 kiloton nuclear rockets, which had the explosive power of about one-tenth of the bombs used over Hiroshima and Nagasaki during World War II. These rockets could be fired at incoming Soviet bombers over the Pacific Ocean as the enemy aircraft flew toward our country.

After launching the Genie rocket, it was mandatory that we execute an escape maneuver to prevent ourselves from being injured by the blast and radiation of the nuclear explosion. We were particularly concerned about the blast effect at night when our vision was extremely important for getting the aircraft back to our home base. We believed that our vision would be lost if we looked in the direction of the blast at the moment of explosion.

The junior pilots asked our commander and tactics instructors how we could possibly return to land with damaged eyesight in both eyes. The senior officers agreed that we would have great difficulty flying our flight instruments at night with degraded eyesight.

After some discussion they came up with a very novel and inexpensive solution: a standard civilian plastic eye patch attached to an elastic string that could be worn over just one eye as we flew the F-101B aircraft. Our helmet and oxygen mask would fit over the eye patch. If a nuclear explosion occurred, we would only lose one eye and could then remove the eye patch from the good eye and fly the aircraft home on that eye. An eye patch was issued to each pilot and stored in a small, zippered red vinyl bag that we could easily insert in our flying jacket and have access to during flight.

About this time Francis Gary Powers was shot down in a Lockheed U-2 aircraft as he was overflying the Soviet Union. The *San Francisco Chronicle* reported that Powers was carrying a suicide kit, which he could use if he were captured. To impress our wives and girlfriends of the danger and responsibility the country had bestowed on junior aviators, we informed them that we had also been given a suicide kit, which we carried in a small zippered red bag. We had found a novel use for the eye patch.

Junior aviators also spent a lot of time after flying in the Fighter Room bar in the basement of the Officers Club. One could wear a flying suit there, which was not allowed in the main bar on the club's ground floor. The Fighter Room bar also was a good place to discuss tactics with other pilots while drinking beer.

Soon it became the custom to wear the eye patch over one eye while drinking in the Fighter Room. To an outsider we looked quite strange—prime physical specimens all wearing a patch over one eye.

As could be expected, the squadron commander joined us for beer one evening. To his surprise he found all of the aviators in his command wearing an eye patch. He made the mistake of asking one pilot why everyone was wearing a patch. It was explained to him that all the pilots were drinking heavily and planned to save one eye with which to drive home.

Before long a directive was published in the squadron that required all eye patches to be turned in to the supply office.

Sitting on nuclear alert was becoming routine for me. However, my interests were soon to be changed forever. At 10:33 AM, May 5, 1961, at Cape Canaveral, Florida, a NASA test pilot sat alone in a small, cramped capsule atop a 69-foot-high Redstone rocket that was about to carry him on a journey through the airless darkness of space. Navy Cdr. Alan B. Shepard, thirty-seven, was born at a time when radio was a novelty, television a dream, and space travel a wild fantasy. He had been sealed into the Mercury capsule atop the rocket for more than four hours, waiting as patiently as he knew how for the moment when he would at last blast off into space.

Many of our squadron pilots were in our ready room, gathered around a black-and-white TV set to watch this flight. One was a Navy exchange pilot, Lt. Cdr. Dusty Rhodes. He was kidding the assembled Air Force pilots about the fact that a Navy pilot had been selected for the first space flight. We responded that a Navy pilot was chosen because the landing would be made in water. If the landing were to be made on a runway, they would have chosen an Air Force pilot. We kidded him that Navy pilots couldn't make a smooth landing on a runway, a problem Dusty was having, so they would make all the water landings. But an Air Force pilot would be selected to land on the moon.

Alan Shepard was eleven years my senior. I had not followed the space program very closely, but now the launch was being covered live on television. Back in 1950, as a boy of fifteen, I watched a Saturday matinee movie titled *Destination Moon*. It was Hollywood's first attempt to portray space flight and seemed more like a cartoon to me. Alan Shepard's flight was real. I had flown the F-101B wearing a pressure suit and knew how uncomfortable Shepard must have been lying in the capsule for hours wearing his suit.

With the countdown in its final moments, we watched intensely for the outcome. The United States was facing a challenge from the Soviets in the conquest of space. A few weeks earlier Yuri Gagarin, a Soviet cosmonaut, had staked a place in history by becoming the first man to enter space.

We too were facing a challenge from the Russians if a nuclear war erupted and we were required to intercept and destroy enemy bombers. The space race took on the sense of a confrontation with the Russians without shots actually being fired.

Alan Shepard's flight of about twenty minutes went off without a hitch, and we started to use his expression "A-OK" in our air-to-air communications too. He landed safely and we cheered like the rest of the nation.

Bill Anders watched with not only great interest but with a sense of purpose. Anders wanted to get into the space program and knew the road to NASA in Houston for an Air Force pilot was through the test pilot school at Edwards Air Force Base.

Soon Anders scheduled a T-33 for a cross-country flight to Edwards to meet with Col. Chuck Yeager, commandant of the test pilot school. I flew along, riding the backseat. I was curious to find out more about the world of flight testing.

While Anders talked with Colonel Yeager in his office, I waited in the school scheduling room viewing the operation. The aircraft assigned to the school for student instruction were impressive: T-38 Talons, F-101 Voodoos, F-104 Starfighters, F-106 Delta Darts, and B-57 Canberras.

The place had an air of intensity and excitement. Groups of test pilots were discussing the test results of their just completed flights while others walked out with helmet, flight test card, and G-suit in hand to explore the unknown. The motto of the Air Force Flight Test Center (AFFTC) was *Ad Inexplorata*, Latin for "Toward the Unexplored."

I did not meet Chuck Yeager, but I did recognize one test pilot who was visiting the school. He was a major, tall and thin, who looked as if he were right out of a Hollywood production. From reading *Aviation Week & Space*

Technology magazine, I quickly recognized him as Maj. Robert M. "Bob" White, one of the test pilots who flew the rocket-powered X-15.[1] White, thirty-six, had entered the Army as an eighteen-year-old aviation cadet in November 1942. Following flight school he flew ground attack missions to cut enemy supply lines, as well as carrying out fighter sweeps against the Luftwaffe. He continued this hazardous flying until February 1945, when he was shot down by heavy antiaircraft fire over Germany during his fifty-second combat mission. He was captured and remained a prisoner of war until his prison camp was liberated two months later. After the war he returned to the United States and enrolled as a student at New York University, where he received a bachelor of science degree in electrical engineering in 1951. During the Korean War he was sent to Japan and assigned to the 40th Fighter Squadron as an F-80 pilot and flight commander until the summer of 1953.

White was then transferred to California to attend the test pilot school at Edwards. He graduated with Class 54C in January 1955 and stayed on at Edwards as a test pilot, flying and evaluating advanced models of the F-86K, F-89H, F-102, and the F-105B. He became the deputy chief of Flight Test Operations, and somewhat later he was named assistant chief of the Manned Spacecraft Operations Branch. White was then designated the Air Force's primary pilot for the X-15 program in 1958. While the new plane was undergoing its initial tests, he attended the Air Command and Staff College at Maxwell Air Force Base, Alabama, graduating in 1959. He made his first X-15 flight on April 15, 1960, when the new aircraft was still fitted with interim rocket engines. Four months later, still using the temporary engines, he took the experimental craft to an altitude of 136,000 feet above Rogers Dry Lake. Thus began the series of flights reaching blazing speeds and altitudes that would soon earn the engineer-pilot high public exposure as well as professional acclaim. After the craft's 57,000-pound-thrust YLR-99 engine was installed, he flew it to a speed of 2,275 mph in February 1961, setting an unofficial world speed record. Over the next eight months, he became the first human to fly an aircraft at Mach 4 and then at Mach 5. This amazing rise climaxed on November 9, 1961, when White reached a speed of 4,093 mph. This was 93 mph more than the plane was designed to achieve and made White the first human to fly a winged craft six times faster than the speed of sound. Following this he took the X-15 to a record-setting altitude of 314,750 feet on July 17, 1962, more than fifty-nine miles above the earth's surface. Flying at this altitude also qualified him for astronaut wings, and he became the first of the tiny handful of "winged astronauts" to achieve that coveted status without using a conventional spacecraft.

President John F. Kennedy conferred the most prestigious award in American aviation, the Robert J. Collier Trophy, jointly on White and three of his fellow X-15 pilots: NASA's Joseph Walker, Cdr. Forrest S. Peterson of the U.S. Navy, and North American Aviation test pilot Scott Crossfield. A day later, Air Force Chief of Staff Gen. Curtis E. LeMay awarded White his new rating as a Command Pilot Astronaut. I listened intently to White explain his most recent flight to other test pilots, though at a distance proper for a junior officer.

Bill Anders was encouraged after his talk with Colonel Yeager. His U.S. Naval Academy education in engineering and his superior grades were real pluses for him. He had fighter experience in both the F-89 and F-101 aircraft and more than the fifteen hundred flying hours considered the minimum by the school to make an application. Colonel Yeager recommended that Anders obtain at least one hundred flying hours of Century Series aircraft experience, be prepared to take a five-day rigorous physical at Brooks Air Force Base, Texas, and obtain a master's degree in a scientific field.

As Anders taxied our T-33 to the Edwards 15,000-foot runway for our return flight to Hamilton, I realized I too was feeling an intense desire to be a part of this new flying frontier. From knowing very little about flight testing only a few hours earlier, I now was convinced that test flying was going to be my life's occupation.

It was going to be much more difficult for me to qualify for admission to the test pilot school than it would be for Bill Anders. My college degree was in chemistry, a science, not an engineering field. My rank was first lieutenant; I was a junior officer with only a thousand hours of flying time. I had been spending many spare hours drinking beer at the Fighter Room bar and had no exercise or physical-fitness training since attending flying school three years earlier. It seemed to me it would be a long shot for me to qualify for test pilot school.

Bill Anders applied to enter a program for a master's degree in nuclear engineering at the Air Force Institute of Technology (AFIT) at Wright-Patterson Air Force Base, Ohio. He was accepted for the program and departed Hamilton within a few months. When Anders set his mind on a goal, he accomplished it. I have never met anyone before or since who had the intelligence, perseverance, and savvy to analyze his strengths and weaknesses and prepare a personal plan of action. I knew great adventure and accomplishments lay ahead for him.

I analyzed my strengths and weaknesses as Anders had done and came up with a plan. First on the list was to become physically fit. I had inherited good health from my parents and always passed the annual Air Force flight physical exam with flying colors. Winning the 100- and 200-yard dash at a track meet in flight school at Bainbridge Air Base was impressive, especially since I had not trained for the event. But the track meet was four years earlier; I immediately needed to start an exercise program. As is the custom in most organizations, both military and civilian, there is a sport that is popular at the time. At Hamilton, it was handball.

Of all places, it was on the handball court that I met Capt. Ralph C. Rich, a test pilot. Rich, thirty-two, a short, stocky crew-cut blond and 1953 U.S. Military Academy at West Point graduate, was a senior captain who had attended test pilot school back in 1958. He was only a first lieutenant when he attended the school, quite rare in those days. From test pilot school Rich transferred to McClellan Air Force Base in the Sacramento area and tested aircraft that had recently come out of depot-level modification.

At Hamilton Rich was a flight-line maintenance officer in charge of all the crew chiefs who serviced our aircraft. The crew chiefs worked directly with pilots as we preflighted our aircraft, started the engines, and taxied out to fly. The crew chiefs were also on the ramp to recover us on our return from flying.

Ralph Rich flew both the T-33 and F-101B on flights known as functional check flights (FCF). After maintenance has been performed on an aircraft, an FCF pilot was required to fly a mission to make certain that the repair or inspection had been completed and the aircraft was safe for squadron pilots to fly. It required extensive knowledge of aircraft internal systems, hand recording data on engine and airframe performance during flight, and the ability to communicate with the maintenance ground crew that performed the repairs.

My physical fitness improved dramatically on the handball court in only a few months. Soon I was known as the one to beat and attracted all comers. Rich was easy to defeat at singles, as his short, stubby legs just couldn't cover the far corners of the court. We did make a good twosome, though, and could hold our own at doubles.

Rich was disgruntled as an aircraft maintenance officer and looked forward to our hours on the court as a diversion. I sensed he was unhappy with his job, but he had not been selected for better flight-testing assignments. He had high expectations for himself that were not being met.

We became good friends and before long he asked me whether I wanted to get checked out as a maintenance test pilot flying FCFs in the T-33. I jumped at the opportunity, seeing it as the second step in my goal of acquiring more hours of flying to qualify for test pilot school. Maintenance test flying would be a step up for me, though it was not for Rich. Although maintenance test flight was not experimental testing Edwards-style, it was entry level into the world I was seeking.

Rich thoroughly briefed me on the procedures and techniques required to accomplish a safe FCF flight. Maintenance test flying was more dangerous than our standard squadron missions, and improper actions could be deadly.

Before long I felt comfortable in being able to accurately fly the T-33 through the entire FCF routine. For the first time in an aircraft, I wrote information other than navigation data on a flight card. I stalled the aircraft with either the landing gear and flaps extended or retracted, recorded the exact speed and altitude, looked for unusual flight characteristics, and explained all of the characteristics to the mechanics on the ground. FCF flights required faster reactions, good memory, and the ability to predict what might be mechanically wrong with an aircraft.

In flight school, spins in the T-33 were a prohibited maneuver. A spin is a complex series of aircraft gyrations, usually resulting from improper pilot control reaction to a stall. In a spin the aircraft violently tumbles in all directions as it rapidly loses altitude. A pilot would not put an aircraft in a spin deliberately. The flight manual, published by the Air Force, was the pilot's bible. The manual described and explained all the aircraft systems, depicted normal and emergency procedures, and had pages and pages of performance graphs. Hidden in a section titled "Unusual Characteristics" was a description of a spin and recovery. The description most likely was written years earlier by an Edwards test pilot during the early stages of the T-33 flight test program. I studied the spin recovery instructions and memorized the recommended procedures. Spins were not required during an FCF. Should I deliberately spin the aircraft anyway? Spinning an aircraft would increase my self-confidence and prove to me that I could handle greater challenges.

The spin recovery procedures worked like a charm. Soon I spun every aircraft scheduled for an FCF. Each responded slightly differently, but now I had enough experience to easily effect recovery no matter how the plane fell through the sky. I developed an added feeling of self-confidence.

On one occasion I tested an aircraft in which the flight controls were badly out of rigging. During the stall series, the aircraft fell off into a spin. Without a moment's hesitation I applied recovery controls, stopped the spin, and safely brought the aircraft home. My study and experimentation had prepared me for the unexpected. My flight hours were steadily increasing; I was ready for more opportunity.

My flying time finally totaled fifteen hundred hours. My application for test pilot school had been prepared for months now. All I needed was to insert the magic number: 1,500.

Gradually I accumulated more flying time and experience in the F-101B. Flight operations were getting to be routine, and I was even beginning to get bored with spending days and days on alert. By late 1962, our sister squadron, the 83rd Fighter Interceptor Squadron, was also flying the F-101B. They kept aircraft on nuclear alert twenty-four hours a day and seven days a week, just like our squadron.

Regulations were very strict concerning the inspection of aircraft with nuclear weapons on board. While the pilot preflighted his cockpit, the RIO would stand at the top of the tall entrance ladder hooked to the side of the aircraft, gun at the ready, making sure the pilot made no mistake that might damage the rocket or cause a nuclear accident. Likewise, the pilot observed the RIO preflight his cockpit. If the pilot didn't follow the Air Force–approved checklist, the RIO was authorized to pull his gun out of its holster, aim it at the pilot, and call for the Air Police to take him into custody. The pilot was authorized to do the same when the RIO preflighted his cockpit. All cockpit switches that armed the nuclear weapon were safety wired to the "Off" position. Electrical power could not be applied to the aircraft at any time during this process, and several air policemen surrounded this whole activity, also armed with guns.

Government rules at the time stipulated that for safety considerations no Air Force nuclear-equipped interceptor would fly over the continental United States unless we were at war with the Soviet Union. Complicated authentication codes and procedures were in place to prevent accidental flight.

Because we had an hour to get airborne if scrambled, the aircrew could go to the Officers Club and meet their families for dinner. Our flight commander carried a portable communications radio, and all aircrew stayed close together. We also spent many hours while on alert in the flight simulator and base gym.

In October 1962, the two fighter squadrons had one of the wildest parties ever at the Officers Club. Alcohol flowed freely, and both squadrons were anxious to outdo each other. Drinks were thrown on people until most of the dancing couples looked like swimmers. Then someone opened a pillow and doused the dancers with feathers. The dancers didn't miss a beat. Now dancing on broken bar glasses with feathers sticking onto their suits and party dresses, they kept going into the early morning hours. It was a party to end all parties.

The next day our flight was on alert. It was expected that the alert crew would assist in cleaning up the club after the wild night. We slowly went through the motions, knowing this was only the start of our three-day alert cycle.

Suddenly a command on the radio ordered us to our planes. We drove in a couple of Jeeps to our aircraft and waited for further instructions. And then we waited longer. The military has always been a hurry-up-and-wait type of organization. Rumor had it that we were going to have a no-notice evaluation by our higher headquarters. If that were the case, headquarters would send a pilot to our base with a written order to our commander. The order would temporarily relieve us of our nuclear-alert requirements. The F-101Bs would be off-loaded of nuclear weapons, and the aircrew tested with written examinations and simulator evaluations. Aircrew assigned to each alert aircraft would then fly practice intercepts against Air Force targets that simulated the Soviet enemy to determine how the aircrew would fare if it were a real war. It was critical for our squadron to score high on these no-notice evaluations.

A box lunch was brought to us as we stood by our aircraft. We had been standing near our F-101Bs for nearly six hours. With nothing else to do, I began a conversation with one of the air policemen standing guard.

"What would you do if we got scrambled?" I asked.

"Sir, I would hold my position," he exclaimed.

With the worldly view of a first lieutenant, I explained to him, "If we get scrambled with these nukes onboard our aircraft, we are definitely at war! The Bay Area is a prime target area; Russian nukes will soon be exploding on San Francisco. We can't possibly stop every incoming Soviet aircraft. I can guarantee you the entire Bay Area would be in ashes."

To emphasize, I added, "If we scramble, your security job is over. Personally, I would get the hell out of here! Head northwest up in the hills to get away from the radiation and fallout."

About that time, the phone at the alert console rang. Our senior officer, a captain, answered.

The deputy wing commander spoke, "Get those planes airborne!"

The captain repeated the words for all of us to hear. Eyes widened and lips took on a thinness of extreme concern. The captain asked the commander to verbally pass on to him the required secret authentication codes.

The commander repeated emphatically, "Captain, I said get those planes airborne!"

Codes or no codes, it was clear the order was to be executed. For a second or so we all held our positions, unable to resolve the contradiction. As in the movie *It's A Mad, Mad, Mad, Mad World*, we waited for someone to make the first move. The tension mounted. Then simultaneously, as in the movie, we each ran to our aircraft. I couldn't believe we were going to fly the F-101B with nuclear weapons loaded.

All the pilots started their engines and we taxied toward the runway. I noticed the 83rd squadron pilots were also taxiing, but toward the opposite end of the same runway we were planning to use. At this time, we were definitely going to fly, unless the deputy wing commander called us back at the last minute.

"I'll bet this is still part of an exercise and the maintenance personnel have off-loaded the nukes while we were cleaning the Officers Club," I thought.

The control tower operator boomed over the radio, "All aircraft cleared for simultaneous takeoff, each maintain the right side of the runway."

Pilots in our 84th squadron pulled out onto Runway 30 in single file and applied afterburner power to start takeoff. The 83rd squadron pilots did the same. The 83rd pilots were taking off on Runway 12, the opposing end of the Runway 30 we were using. The two squadrons' aircraft were headed directly at each other, separated laterally by only a few feet. Not even the Air Force Thunderbird Demonstration Team would ever be approved for such a risky maneuver. Now I knew we were at war!

One of the senior captains in this flight had attended the Interceptor Weapons Instructor School at Tyndall Air Force Base, Florida. He was now airborne in our flight acting as our deputy flight leader. He was an absolute stickler for regulations and details. I called him Pain-in-the-Butt since he discussed every tactic and maneuver like a lawyer. We clashed on most flying subjects. He was legally correct with his statements but living in a dream world, I thought. I believed a real war would never be fought using his strict approach to the military regulations.

As we flew northwest approaching the Pacific coastline, I kept looking back at San Francisco. Any moment now incoming Soviet intercontinental ballistic missiles would be exploding, and all our lives would be changed forever. I had left a wife and two-and-a-half-year-old son back home whom I might never see again. I turned around in the cockpit and looked back at Hamilton Air Force Base for what I thought could be the last time. Far off on the horizon the beautiful red Golden Gate Bridge was still shining brightly in the late afternoon sun. I wondered what had happened to the air policeman I had been talking to. Did he take my advice?

Off to my side and slightly below me, I saw an F-101B with its landing gear still extended.

"One of our pilots must be having a gear problem," I called over the UHF radio.

I then noticed it was Pain-in-the-Butt, our deputy flight leader. He radioed that he couldn't get his nose landing gear to retract. I joined in close formation only about ten feet off his left wing and quickly observed what had caused his problem. He had failed to remove the nose landing gear safety pin prior to takeoff. His nose landing gear would now remain extended for the duration of the flight.

To make things even worse, because the nuclear rockets were loaded internal to the aircraft and directly behind the nose gear, his entire radar and weapon system was automatically deactivated so as not to destroy his aircraft if his Genie rockets were launched. Here we were on the biggest potential combat mission that any of us could ever imagine in our wildest dreams, and Pain-in-the-Butt was already out of the battle.

He did have one possible use: Our tactics called for us to ram (fly directly into an enemy aircraft, make physical contact, and knock it down) if we were out of weapons or had a disabled weapons system. I wondered whether old Pain-in-the-Butt would follow his cherished Air Force regulations and actually ram an enemy aircraft.

We orbited over the Pacific Ocean for an hour or so. Nothing happened except for radio communication between Pain-in-the-Butt and the ground radar controllers. The war now seemed to be centered on his problem. No one was looking for the enemy.

"Where the hell are the Soviets?" I wanted to know.

This would have been a good time to make a caustic remark over the radio to Pain-in-the-Butt about the major mistake he had made. Instead I held my tongue; maybe I was developing the maturity to become a true fighter pilot.

Low fuel level was slowly becoming a critical factor as we continued to orbit. Around and around we turned at 35,000 feet, the city of San Francisco just a speck on the horizon. Suddenly, just like the order to scramble, we were given an order to fly north and land at a civilian field in Oregon.

As soon as I landed in Oregon, I opened an inspection plate on the underside of my plane. Sure enough, the two nuclear rockets were really on board. We were not playing games, nor was this the no-notice inspection we were all expecting.

Pilots and RIOs gathered around an old black-and-white TV set as President John F. Kennedy spoke to our nation about the crisis. We learned, with the rest of our country, about the Soviet missiles being placed in Cuba. The entire Air Force was on maximum alert. Our Air Force Air Defense Command commanders wanted to disperse their fighter aircraft away from the missile threat in Cuba. Russian bombers were not flying across the Pacific Ocean to the United States . . . at least not yet.

Once again we were back on ground nuke alert. Days and days went by until our wives could package clean flying suits and underwear and have them shipped to us on a military transport. Now time really dragged on and on. No flying. No parties. No gym. No war.

Slowly, the federal government resolved the Cuban Missile Crisis and our squadron commander made plans to return the F-101B aircraft and nuclear rockets to Hamilton. As a safety precaution, he decided to have the rockets deactivated internally; place heavy-duty, nonbreakable safety wire on all cockpit weapon switches; and ferry each aircraft one at a time back to our home base so that there was no possibility of collision. The war was over—and not a shot had been fired. But the crisis was the closest the United States and the Soviet Union came to an all-out nuclear exchange during the Cold War.

Pain-in-the-Butt had time to remove his nose wheel safety pin before he flew home.

Bill Anders completed his master's degree at the Air Force Institute of Technology with honors. He even had a chance to continue on for a PhD. He didn't think an additional degree would enhance his standing on his test pilot school application, so he turned down further education.

Anders was then assigned to the Special Weapons Center at Kirtland Air Force Base, New Mexico. The center specialized in the development of nuclear weapons; it was a natural choice for someone with his experience and background.

As soon as he got settled in his new job, Anders contacted Colonel Yeager again to get the latest information before he sent in his application to test pilot school. Yeager told Anders circumstances had changed at the school and they were no longer looking for pilots with advanced degrees. Yeager said he could teach his pilots all they needed to know. Now the school wanted pilots with more Century Series flying time—the more time, the better.

Anders's application was considered as "qualified" by the test pilot school, but not "highly qualified," meaning he had only a slim probability of being selected. With the tremendous publicity being generated for the NASA space program, hundreds of applications had been flooding into the school. Anders had taken the path Yeager suggested and it turned out to be a dead end. Irritated, he told Yeager, "I'll get into the space program without going to your damn school!"

Yeager could not have cared less. A rocket-powered NF-104A Starfighter was being prepared as a space trainer for the school. Yeager was scheduled to attempt a world altitude record in the aircraft. He didn't have time to argue with a candidate trying to qualify for the school. Yeager had no shortage of applicants; he didn't care about Anders. Bill Anders applied directly to NASA to become an astronaut.

Back on the handball court, my physical conditioning leveled off. Spending days and days on alert was becoming a nightmare for me. An assignment to Glasgow Air Force Base, Montana, came with my name on it. Leaving sunny California and returning to the cold wintry weather of the northern states for three to four years was my future if I was not accepted for test pilot school. This new assignment to Montana was on temporary hold pending results of the test pilot selection board. Shortly, I would be moving either north or south—it was out of my hands. For the first time, I considered getting out of military service and becoming a civilian. Jan loved the Bay Area and we felt comfortable living off base in a civilian community.

My flying time in the F-101B was approaching eight hundred hours. I was not high man in the squadron, but my flying time was very respectable for the time I had been stationed at Hamilton. Ralph Rich had a question for me. Would I like to get on military orders to fly functional check flights in the F-101B? At that point, I would have rather flown FCFs all day than ever be on nuclear alert again. Orders were processed and I took on the new task with renewed enthusiasm. With my previous experience flying FCFs on the T-33, the F-101B checkout was a breeze.

On March 28, 1963, I read in the local *Novato Advance* newspaper, "Capt. Ralph C. Rich, 33, maintenance officer for the 78th Fighter Wing at Hamilton AFB, was found dead of a gunshot wound today in his home on the base. Air Police listed the wound as self-inflicted pending further investigation. Captain Rich's wife, Patricia Joan, reported she was awakened by the shot at 2:20 AM, went into the dining room and found the body on the floor. The shot had been fired from a .45-caliber pistol which lay beside the body, the Air Police said. There was no explanation of why the captain might have wished to die. In addition to his wife, he leaves two children, Marisa, 3, and Kerri, 2."[2]

A good friend had taken his own life. He was someone who, in my estimation, traveled the road I now desired and had a lot to live for with a wife and two small children. Evidently Rich felt very differently. Not only did I not suspect Rich was that unhappy with his life, he had not asked me, a close friend, for help. It was as if we had an argument and he had the last word. I was very upset. I asked to meet with the minister of the local church we attended. Suicide is a supremely selfish act that ignores the grief of loved ones left behind. Ralph Rich had taken a part of me with him.

Rich's death was the low point of my three and a half years at Hamilton. Generally life had been good for Jan and me with the birth of our son, Randy, three years earlier, and we had enjoyed personal stability, living in one house the entire time. We were even thinking about adding to our family.

Into the depths of my personal loss and despair over Rich's death came the good news that I had been selected as one of two alternates for Class 63A at the test pilot school. This class would start in July 1963, only a few months away. All of my extra flying had paid off and now only the physical examination at Brooks Air Force Base stood in my way to becoming a test pilot candidate. For the first time since pilot training, I started jogging on the high school track. Round and round I ran at night looking up at the stars.

Early in June 1963, I arrived at Brooks Air Force Base, Texas, to start a five-day, pre-astronaut physical examination. In five days, the medics had time to put a human body through every imaginable test. As an alternate, I knew I had to excel in every way possible.

At Brooks I met Capt. Charles R. "Chuck" Rosburg, another candidate for the school. Rosburg was born and raised in my college state of Iowa. He had flown the Boeing B-47 Stratojet bomber in Strategic Air Command for many years. He also had stood nuclear alert similar to me. Rosburg reminded me of the popular Alley Oop newspaper cartoon caveman character with his broad shoulders and muscular arms. He told me he had spent over half of his military service working out in the gym. He lifted weights and played handball and racquetball. I had never seen a pilot who appeared to be in such good physical condition. The Air Force had chosen an excellent specimen in Rosburg—the competition was going to be formidable.

Each candidate expected to have his sight and hearing tested. We had completed those tests every year on our regular annual flying physical. Nothing could have prepared us for what was to come. An electrocardiogram was taken while we had one hand submersed in ice water, and another was taken while we were hanging from a parachute harness. We spent several hours with a psychologist discussing our relationship with our parents, our use of alcohol, our sex life. It was unclear to me how to beat your competition in these special events.

The sigmoidoscopy examination, or "insertion of the steel eel," as it was then referred to, was the showstopper. The candidate was required to undress and lie on his stomach on a hospital table with his knees on a lower foot pad. The table was then rotated so that the person was upside down, supported only by his shoulders. The physician inserted a four-foot-long aluminum bullwhip with a searchlight, a remote viewing device, and a lasso at the tip, looking for polyps in the large intestine. This procedure was painful, uncomfortable, and embarrassing. I had volunteered for this unique evaluation only because I wanted to get into the test pilot school.

Earlier I heard that candidates would be evaluated during a treadmill test, something I looked forward to. We were required to run on a moving belt while hooked up to medical monitors that measured heart rate, blood pressure, and electrical impulses in the heart. A blood sample was taken every minute. The goal was to run until a person was exhausted or until the medics felt they were in distress, in which case the test was stopped. In this event we were directly compared to see who was in the best physical shape. Rumor had it that a rectal thermometer was used as part of the test.

I was so motivated to do well that I gave my all to the test. The medics reported to me that I had stayed on the treadmill longer than any other pilot. Only a professional decathlete had stayed on for just a few seconds more. I wished the selection would be decided simply on the treadmill results.

As a group, there was one competitive event outside of the medical evaluation. In preparation for an evaluation of our intestinal systems, a radioactive barium compound was consumed. The radioactive material would decay in a short time and the doctor wanted to examine all candidates at close intervals. To empty the lower intestine, all candidates were required to give themselves an enema. A package with a Fleet enema was handed to us. Wearing only hospital gowns, we marched down the hall as a group to accomplish the task. A Fleet enema was a plastic bottle, the size of a ketchup container and filled with a clear liquid, which was to be placed where the sun never shines and was absolutely guaranteed to be successful. Without a word being said, we knew the real competition was at hand. Let the games begin.

Every pilot entered a private bathroom stall, a separate world, but one that was only a few feet from everyone else. No one said a word—this test was serious business. In our own way, we would determine the top candidate. Simply squirt the chemical in and start the clock. The last man sitting, so to speak, was the winner. Time passed slowly. My stomach started to gurgle. I was sure no one else had practiced this special event; we were all entering our personal unknown.

I recalled the Edwards Flight Test Center motto, "Toward the Unexplored." But I didn't think this test was the "unexplored" the flight test center had in mind!

My eyes began to water. Moans could be heard, but the person responsible was not identified. Just hold on, hold on. Suddenly an explosion occurred. Who let go I didn't know, and by then I didn't care. All hell broke loose—it was every man for himself. In rapid fire, fumes and laughter filled the room. We all had lost and we all had won.

Capt. James E. "Jim" Taylor was the last to emerge from this private war. I had admired many men in my life: flight instructors from pilot school, a Korean War ace I met in college, and the minister who helped me recover from Ralph Rich's death, but Jim Taylor showed test pilot coolness under pressure. He got my unofficial vote and I hoped I would get a chance to fly with him in the future. This unintended medical test was one I was happy the medics never observed or recorded.

Before returning to California, the doctors informed me I had passed every phase of the physical exam. One prime candidate had failed the exam. Either the other alternate candidate or I would be selected as his replacement. It was a 50-50 chance for me now.

Back at Hamilton I waited and checked into our squadron administration office every few hours for results. In a few days, the squadron received a military Teletype message addressed to me. The test pilot school had selected the other alternate. The message indicated that my chances of being selected for the next class looked promising, but it offered no guarantees. Bad news.

Jan had good news to report—she was expecting again.

I returned to the handball court, nuclear alert, and an uncertain flying and military future.

During a handball game, I was in the front of the court and hit the ball very hard. I waited for my opponent to return the volley. Seconds passed with no return. I turned toward the back of the court to retrieve the ball. Suddenly the ball smashed directly into my right eye. The resulting swelling soon completely closed my eyelid. I considered whether to contact our flight surgeon. If I did, the injury would be recorded on my medical record. Knowing how detailed the Brooks physical exam had been, this injury could disqualify me for test pilot school. But since I could not see out of the right eye, I had no choice but to contact the flight surgeon. I was medically grounded.

In the early fall of 1963, Bill Anders was selected by NASA to become an astronaut. He told me that NASA liked his educational credentials and had invited him to Houston for an interview. He found that he would be competing against other Air Force pilots who had graduated from the test pilot school and had flight tested for several years. Anders had thought long and hard about the best strategy to use in the interview. His strong suit was his knowledge of radiation in space. Anders told me he made a case to the interview board that radiation was a great hazard to astronauts flying a capsule in space. He offered to work with NASA engineers as an astronaut to develop radiation shielding. The board agreed with his concern and with his concept of how to correct the hazard. This hazard was a little embellished, Anders knew, but the board didn't. Bill Anders was selected as a NASA astronaut. He had made an end run around Colonel Yeager and his test pilot school. Anders was planning to go to the moon.

Another military Teletype message came into our squadron administration office. I was a prime candidate for the test pilot school Class 64A starting in January 1964. The message required my return to Brooks for another five-day physical exam. The swelling had gone down from my eye injury and my sight was normal except for an enlarged pupil. I wondered what kind of strategy Bill Anders would have thought of under these circumstances.

Arriving at Brooks in November 1963, I met the other pilot candidates for 64A. Since I had completed the full five-day physical only six months earlier, this new group considered me to be the "old guy." They all wanted to know what kind of physical tests would be performed. I used words some had never heard before: Fleet enema and sigmoidoscopy.

The Brooks medics informed me that I would be required to take only selected parts of the physical exam again. The eye test was not required.

"How could someone's eyesight change in just a few months?" the medics asked.

It was a question I didn't answer. Anders would have approved of my response. The pressure was off. However, the treadmill test was required again. This time, due to either overconfidence or lack of the desire that I had only a few months earlier, I turned in a lackluster performance. It showed me how important motivation was in physical achievement. A lesson learned.

The second treadmill test wasn't decisive, however. I was selected for Class 64A. It was time to get ready for a move after four years of living in one house.

Some of the test pilots at Edwards bought an extra old car to drive back and forth from the housing area to the flight line. In the 1950s, a Model A Ford was the choice of fighter test pilots. Following their example as best I could, I bought a 1950 blue Ford convertible for seventy-five dollars to use as a commuter car. The antique auto looked perfect for the part of a fighter test pilot. Fighter pilots flew fast—but they drove slowly on the way to work!

I needed a large desk to use for the many hours I would be reading and writing test reports. Jan and I found a used wooden desk at a garage sale for five dollars. I also needed to take military leave (vacation in the civilian world) before the move, as time would not be available in the following year. I scheduled two weeks of vacation the last part of November and stayed at home.

During the third week of November 1963, I was refinishing the wooden desk in our garage. I was listening to the radio when the program was interrupted by a special bulletin, "President John F. Kennedy has been shot and killed in Dallas, Texas."

Jan and I both remember the exact moment we heard the terrible message from Walter Cronkite on CBS News. We grieved and wondered how President Kennedy's death would affect the space program. All government agencies went into official mourning. My going-away party at the Officers Club was canceled. It was just as well: I wasn't drinking beer anymore because I was still involved in the intense physical training program that I had devised, and my eye patch had been turned in to supply.

Fast-moving events were overtaking our nation during Christmas of 1963. A nasty little skirmish in Vietnam was making small headlines in the *San Francisco Chronicle*, and students at the University of California across the bay in Berkeley were protesting something, God knows what.

After Christmas, Jan and I packed up and moved on. The Golden Gate Bridge would have to be defended by a new generation of pilots. It was still shining as brightly as the day Bill Anders and I had crossed it four years earlier. Jan was pregnant for this move, just as she was during our last move. I drove into Hamilton in a 1958 blue MGA roadster and drove out in a 1950 blue Ford convertible. It might not have seemed like progress, but I was going to be a test pilot.

Chapter 6

Charm School

Col. Chuck Yeager, the grand old man of supersonic fame, sat on the high throne of flight test when I arrived at Edwards Air Force Base in late December 1963 to start test pilot school. Our class was composed of sixteen pilots, ten from the Air Force, two from the Navy, two from NASA, and one each from Canada and the Netherlands. We all had been flying from six to nine years and had over 1,500 hours in aircraft. Three of the pilots were bachelors. The rest were married, and most had several children. My wife delivered our second son, Scott Steven Marrett, just three months into my training. For the bachelors it was slim pickings on base. Edwards was as remote from civilization as some foreign military assignments, and few single women lived around the base.

Yeager became commandant of the school two years before I arrived and was guiding the Air Force into the space age in competition with NASA. The Air Force had its first manned orbital space program in the Boeing X-20 Dyna-Soar. The X-20 was a single-piloted space vehicle, to be launched into polar orbit from Vandenberg Air Force Base near Lompoc, California, with plans to conduct surveillance of the entire globe as the earth rotated under the X-20's pole-to-pole orbit. After orbiting the earth, the vehicle would reenter the earth's atmosphere and a test pilot would fly a dead-stick (nonpowered) landing onto the Rogers Dry Lake similar to the X-1, X-2, and X-15. Dyna-Soar would make its first flight in 1966, just three years away.

"Pilots won't have to brush monkey turds off their seat when they fly the Dyna-Soar like the NASA astronauts in Houston," Yeager proclaimed, referring to the use of monkeys in the space program.

All Dyna-Soar test pilots would be handpicked by the Air Force and fly their space capsule from liftoff to landing on Rogers Dry Lake.

Yeager was instrumental in changing the curriculum of the test pilot school from one that taught aircraft testing only to one that now included space-flight training. The name of the school was also changed—to Aerospace Research Pilot School (ARPS), though it was more commonly referred to as Yeager's Charm School. Yeager had an unlimited credit card account with the Air Force. If he needed equipment for the school, he got it. If he needed transportation for military business or a hunting trip, he got it.

"Follow me" seemed to be his motto. "I'll put the Air Force in space." Yeager still had the golden touch.

Training for student astronauts included flying a space mission simulator, a flight in a Boeing C-135 aircraft to practice zero-G profiles, and high-G space reentry experience in a human centrifuge at Brooks Air Force Base, Texas. To give his students a real taste of space, Yeager contracted with Lockheed Aircraft to modify three production F-104s for high-altitude flight, adding a rocket in the tail and small directional thrusters in the nose and wingtips. The aircraft, now called the Lockheed NF-104, would take a student to over 100,000 feet in altitude.

The production Lockheed F-104 Starfighter was the kind of aircraft you put on like a glove. The cockpit was small and compact but still comfortable. Lockheed press releases called it "The Missile with a Man in It." It had a long pointed nose and straight stubby wings canted downward that were so thin one wondered how it could fly. The leading edge was so sharp a pilot could cut his hand on it during preflight. He wore what looked like cowboy spurs on his flight boots. These were attached to cables at the base of his seat, so that if he ejected, the cables would automatically pull his feet up against the ejection seat and he wouldn't injure his legs or feet. Most pilots put the spurs on just before they boarded the F-104 and took them off immediately after deplaning. Others really liked their spurs.

Years earlier, when I was attending basic flying school at Webb Air Force Base and flying the T-33, I saw an Air Force colonel who had just landed in an F-104. He walked into base operations wearing an orange flying suit and dress military hat with scrambled eggs on the bill signifying his high military rank. His Starfighter spurs were still attached to his flying boots, clanking as he walked. He was really calling attention to himself. It was an impressive sight to me as a second lieutenant student flight officer. I decided then that I wanted to fly the Starfighter someday, chasing an even faster whistle.

The NF-104 was a modified F-104 with a 6,000-pound-thrust Rocketdyne rocket engine attached to the tail; it used JP-4 and hydrogen peroxide for fuel. This combination permitted flights from 106,000 to 120,000 feet, depending on the pilot's climb angle. Equipped with the same reaction control system (RCS) used in the X-15, a student astronaut could control the NF-104 on a zero-G trajectory through the high altitude of near space for about eighty seconds. This training was accomplished in a full-pressure space suit, fully inflated due to lack of cockpit pressurization when the J79 engine was shut down during the zoom climb. Flying a zoom flight in the NF-104 would give the student astronaut confidence he could fly to the edge of space and safely return.

Yeager hadn't tried to break a record in the skies over Edwards since 1953, when he had set a new speed record of Mach 2.4 in the Bell X-1A. Whoever was to push the sleek NF-104 to optimum performance was certain to set a new world record for altitude achieved by a ship taking off under its own power. The Soviets had set the current record, 113,890 feet, in 1961 with the E-66A, a delta-winged fighter plane. Even though the X-1, X-2, and X-15 had flown higher, they had to be carried aloft by a Boeing B-29, B-50, or B-52 before being dropped and then rocketed to record-breaking altitudes.

Lockheed modified three production F-104s and began shakedown flights with company test pilot Jack Woodman at the controls in the summer of 1963. After a few months, the test program was turned over to Maj. Robert W. "Smitty" Smith, a test pilot at the Air Force Flight Test Center flying out of the Fighter Branch of Flight Test Operations. A year later, after I graduated from the test pilot school and was assigned to the Fighter Branch, I met Smitty and did a little off-the-record practice aerial dogfighting against him. Smitty would disable the aircraft's safety system, which was used to prevent loss of aircraft control at high angle of attack and high Gs, and maneuver the F-104 as close to its limits as was humanly possible. You couldn't beat Smitty while flying another Starfighter against him.

During the NF-104 flight test program, Smitty reached a maximum altitude of 120,800 feet on one zoom, although it was not a world record because it was a test flight and the FAI (Federation Aeronautique Internationale) monitors were not in place. Optimum climb angle for the aircraft turned out to be between 65 and 70 degrees. Combined with a 14-degree seat cant and 5-degree angle of attack, this left the pilot reclined at about an 85-degree angle from the straight up. These high angles prevented visual reference to the ground, so all zoom maneuvers were made on instruments just like the X-15. On one test flight, Smitty climbed the aircraft at a pitch angle of 85 degrees to see if this would improve performance, but he lost control and tumbled, going over the top upside down. The aircraft entered a spin, but he recovered. Smitty was absolutely fearless!

Just like the old days at Edwards, Yeager was going after a world record. Time was short; Yeager wanted the record set by mid-December, the 60th anniversary of powered flight. Time was quickly counting down, so the planned sequence of build-up flights was canceled. According to Smith, Yeager was in a hurry to set the record and didn't wait for the preceding flight test data to be evaluated.

Yeager simply said, "If Smith can fly it, I can."

He had made a near-fatal mistake.

Yeager had taken NF-104, serial number 56-0762, up for three checkout flights, getting the Yeager feel for the Starfighter. On December 10, 1963, he was scheduled to fly two practice zoom flights in preparation for the all-out record attempt the next day. On the morning flight he reached a peak altitude of only 108,700 feet, but Yeager knew the Starfighter could be taken much higher.

On the afternoon flight, Colonel Yeager's test profile called for him to accelerate to Mach 1.7 at 37,000 feet, fire the rocket engine to accelerate further to Mach 2.2 at 40,000 feet, and then climb at 60 degrees. Yeager followed the test plan, but as the aircraft passed through 70,000 feet, ground control called on the UHF radio that he had less than the desired angle of climb. Yeager used the reaction controls to get back up on his flight path. He had previously used this technique without a problem. However, on this flight he was at a lower altitude and had a higher dynamic pressure.

When Yeager attempted to lower the nose, he found he had lost both aerodynamic and reaction controls. Soon he fell into a spin and gyrated in all directions. Nothing he tried stopped the spin. Only a mile above the desert floor and descending like a manhole cover, he ejected.

As Yeager's parachute opened, he was struck in the face by the base of his rocket seat, breaking his visor. Burning residue from the rocket entered his helmet. With breathing oxygen flowing to the helmet, an intense flame started to fry his neck and face inside his space suit. As he descended in the parachute, Yeager removed his left glove and used his hand to try to put out the fire around his nose and mouth. He charred two fingers and a thumb.

The aircraft continued to spin and contacted the ground in a very flat attitude. It "augured in," to use an old Yeager phrase. In just a few minutes an Edwards helicopter piloted by Maj. Phil Neale, with Major Smith and a flight surgeon onboard, arrived on scene and transported Yeager to the base hospital where his pressure suit was cut off to prevent aggravating the burns on his face and hand. Colonel Yeager sustained second-degree burns on the left side of his face and neck and on his left thumb and two fingers, and he was cut on his upper-left eyelid.

After an extensive investigation, an Air Force accident board concluded that Colonel Yeager had contributed to the cause of the accident by "purposely exceeding the recommended angle of attack in order to attain a higher altitude. He had also exceeded the recommended angle of attack on his previous flight, but probably had been able to get the nose down because of a lower dynamic pressure." In addition, the board stated that "the primary cause of the accident was that the aircraft could not be recovered from a spin that resulted from excessive angle of attack, and lack of aircraft response."

Major Smith stated on his Web site,

> Colonel Yeager lost control of the airplane during ascent before he could achieve the prime mission of space flight training, causing an unrecoverable flat spin, which resulted in loss of the NF-104 aircraft and his own serious injury. Detailed data on all major systems of the airplane during the accident demonstrated that all systems were functioning properly and that the aircraft had no failures whatsoever, and no mitigating contributions to the accident. There were no other contributing factors. The aircraft had a higher than normal risk factor, as clearly understood in the intent of design and procurement for the special purpose of extremely sophisticated flight into near space regions and astronaut training at affordable costs, which it proved capable of doing on numerous full zoom flights. Its systems had been proven fully capable and sufficient to perform the mission, whenever it was flown within the prescribed profile and by pilots trained for both aero and space control. Yeager, an outstanding test pilot, however, did not have the technical background and training for space.[1]

Smith also said,

> A flight test engineer told me that the stability derivatives from his flights had not been reduced until after Yeager flew. The engineer tried to get Yeager to wait until they were done and he could train in the simulator but he refused. That's the real reason that Yeager lost the airplane and almost lost his life. The real tragedy is that when Yeager failed to recover, the consensus was that the test school could not fly such a dangerous aircraft and the project was cancelled, thereby depriving the school and its students of a tool that would have allowed real cutting edge training and development of some crackerjack test pilots.[2]

Yeager had been in such a hurry to set the world altitude record that he had not taken advice from Major Smith and a flight test engineer before he was back in the saddle with his spurs on raring to go again. The horse that bucked him off before his first supersonic flight in the X-1 rocketship sixteen years earlier kicked him this time!

The loss of one of the three NF-104s was not the only bad news that day. Secretary of Defense Robert S. McNamara announced the cancellation of the X-20 Dyna-Soar program. The Air Force was now without a manned space program and Yeager was wrapped in bandages. The Air Force put a hold on Yeager's credit card.

Jan was delighted when I was selected to attend the test pilot school. However, she was concerned about the number of fatalities of pilots during my previous four years in the 84th Fighter Interceptor Squadron at Hamilton and suspected that flight testing would be even more dangerous. She did not want our children to grow up without their father. When she asked me about the dangers of test flying, I found it a difficult question to answer. I explained that I would get the best training in the world, that test aircraft were maintained to a higher level than fighter squadron aircraft, and that my experience in flying maintenance test flights would prepare me for the greater challenges ahead. I wasn't sure my answers were entirely correct, but she seemed to accept my explanation and let it pass.

The school planned an open house for new students and their wives. This was an opportunity for the families to meet our instructors, see the classrooms and simulators, and view the test aircraft in the school hangar and on the flight line.

We waited in the school auditorium for Colonel Yeager to make his grand entrance. He finally arrived wearing his dress blue uniform with row upon row of ribbons awarded for flying combat in World War II and for his historic flight test accomplishments. However, he also wore a large white bandage around his neck, and his left arm was in a sling to immobilize his burned thumb and finger. His bruised and bandaged appearance was not comforting to my wife. It was obvious she sensed that flight test could be a very deadly and dangerous business if the premier Air Force test pilot was all banged up.

As if seeing Colonel Yeager in bandages was not enough, the tour of the test pilot school hangar held another surprise. All of the school aircraft had been removed so that the wreckage of Yeager's NF-104 could be placed on the hangar floor for inspection and analysis. The only recognizable parts of the aircraft were the T-tail and part of the rocket engine. I had investigated several aircraft accidents earlier, so the sight of twisted and tangled wreckage was not new to me. But most civilians, and certainly not our wives, had never viewed such destruction. It was a shocking sight. I did not realize it at the time, but this sight of destruction was a sign of times to come with thirty-two of my test pilot friends killed in aircraft crashes over the next twenty-five years.

As a new student in the school, I soon had my first flight in the backseat of the F-104 with Maj. Frank E. Liethen Jr., one of the instructors, as he conducted a functional check flight. An FCF was flown on every military aircraft after major maintenance had been performed. Usually only the most experienced pilots were asked to fly these potentially hazardous flights.

Liethen, thirty-four and a native of Appleton, Wisconsin, attended the U.S. Naval Academy at Annapolis and graduated just as the Korean War ended. Accepting a commission in the Air Force, he completed pilot training at Marana Air Base, Arizona, and Williams Air Force Base, Arizona, and F-86 Sabrejet gunnery training at Nellis Air Force Base, Nevada. After completing the gunnery course he was assigned to the 494th Fighter Bomber Squadron at Chaumont Air Base, France, where he flew the F-86 and F-100 for three years. He returned to the United States to attend the Air Force Institute of Technology at Wright-Patterson Air Force Base, where he received his master of science degree in electronics. For two years, Liethen remained at Wright-Patterson as a fire-control system engineer working in the F-105 Thunderchief Systems Program Office (SPO).

Liethen attended test pilot school from August 1961 to April 1962 and was chosen as the outstanding pilot and overall top student in his class. After a year as a project test pilot in the research and development section of the USAF Fighter Weapons School at Nellis, he returned to Edwards to attend the new space school. Following graduation in December 1963, Liethen became an instructor in the school.

Frank Liethen had a great desire to become an astronaut. However, the Dyna-Soar program was canceled as he graduated from the school, so that program was not available to him. He had applied to NASA for acceptance into their astronaut program but had been turned down because his height exceeded their standards. His last remaining chance for a flight in space was the next-generation Air Force manned space program on the drawing board, the Manned Orbiting Laboratory (MOL). The selection of astronaut candidates for the new program would be announced in late 1965, a year and a half later.

At Hamilton, I had become proficient in flying FCFs in the F-101B. In many respects the F-101B and F-104 were similar. Both had a T-tail and were designed for supersonic operation. The F-101B Voodoo was a twin-engine two-man aircraft with an RIO operating radar; it could fire a MB-1 Genie nuclear rocket. The F-104 had a single General Electric J79 jet engine with afterburner and short-range air-to-air radar to detect enemy targets and then fire the heat-seeking Sidewinder missile. Due to the Starfighter's short range, all of the Air Defense Command squadrons had been deactivated and the planes transferred to Tactical Air Command to be used as fighter-bombers in Vietnam.

Because the F-101 and F-104 were designed with a T-tail, both would pitch-up if maneuvered to an excessive angle of attack. Pitch-up generally resulted in a spin. Both aircraft were difficult and sometimes impossible to recover from a spin. If a pilot was fortunate enough to recover from a spin, the ensuing dive recovery could take up to 10,000 feet in altitude. Pilots did not intentionally pitch-up or spin either the F-101 or F-104; the pilot's flight manual called it a prohibited maneuver.

The F-101 and F-104 each had an electronic safety system to prevent a pilot from entering the pitch-up region. The F-101 had an audible horn that sounded in the pilot's helmet as G increased near the pitch-up boundary. If the aircraft continued to be flown at even a greater angle of attack or G, a pusher physically pushed the control stick forward, reducing the angle of attack. This was a very complex electronic system for the day and required the FCF test pilot to adjust the electronic boundaries in flight.

The F-104 instrument panel displayed an instrument with an angle-of-attack needle. As a warning that the pilot was approaching the pitch-up boundary, the needle would edge into a red area on the display. If the pilot continued to increase angle of attack or G, a stick shaker activated. This system caused the control stick to shake in the pilot's hand and emitted a rattle-like sound similar to that of a rattlesnake.

Flying either a F-101 or F-104 near the pitch-up boundary and adjusting the electronic system not to be overly restrictive to the squadron pilot but not loose enough to be a safety hazard required good pilot feel and a smooth touch.

The Starfighter wasn't much in a dogfight. I felt the flight controls were too heavy and limited Gs were available due to its pitch-up tendency. But it would blast out to Mach 2 very quickly and zoom to high altitude.

As Liethen performed maneuvers in the F-104 from the front seat, I held the control stick ever so lightly in my hand and watched intently as we neared the maneuver boundary. Liethen talked to me on the intercom as he flew the plane, and I watched him like a hawk. He was responsible for the safety of the aircraft, but my butt was in the plane too. I was impressed by Liethen's handling of the Starfighter and looked forward to the latter half of the school curriculum when I would get checked out.

The F-104 had been the first Air Force aircraft to use the General Electric J79 engine. In the early days, the J79 engine was very unreliable and failed in flight on a regular basis. Only four years earlier 1st Lt. Ken Mason, my flying friend at Hamilton, had been killed when the engine quit and he was forced to eject.

Iven Kinchloe, an Edwards' test pilot from the 1950s who had set an altitude world record in the Bell X-2, was killed in the F-104 in 1957 when his engine failed just after takeoff. The Starfighter had a terrible safety record.

"Plan to go vertical," Maj. David H. "Dave" Tittle, one of my flight instructors at the test pilot school, told me.

I tried not to blink or show any indication of apprehension, but I cringed at the thought of flying a jet aircraft straight up. The vertical maneuver required pulling the nose of a Northrop T-38 Talon to a 90-degree climb angle, watching the airspeed bleed off to zero, taking my hand off the control stick, and going along for the ride. It sounded very risky.

Tittle, thirty-five, was from Springfield, Ohio, and had flown one hundred night combat missions in the A-26 Invader during the Korean War. He was awarded the Distinguished Flying Cross and three Air Medals. After graduating from test pilot school in 1955, Tittle flew test missions at Wright-Patterson Air Force Base for five years before getting a master's degree from the University of Oklahoma. He was married with two children.

In the previous four years I had flown eight hundred flight hours in the F-101B Voodoo. If I had flown a vertical maneuver in the T-tailed Voodoo, the plane would have pitched up, stalled, and fallen into a spin. Spins were usually irrecoverable—the only option then would be to eject. Avoiding a vertical maneuver had been my goal for the last four years. But now I was training to be a test pilot and venturing into the unknown. Going vertical was part of the school curriculum in checking out in the T-38.

I had eyed the two-seated Talon enviously since I arrived at Edwards. Sleek and needle-nosed, with thin, wedge-shaped high-speed wings, it had many of the earmarks of a fighter. In fact, a single-seated tactical version of the Talon, the F-5A Freedom Fighter, was being tested just down the flight line at the Fighter Branch of the Air Force Flight Test Center.

The Talon was a little over 46 feet long, had a wingspan a bit over 25 feet and an overall height to the top of the vertical stabilizer of 12 feet 11 inches. Fully fueled, the Talon's gross weight on takeoff was just under 12,000 pounds—very light for a supersonic-class aircraft. It had two General Electric J85-5 turbojet engines with afterburners, which produced a total thrust of nearly 8,000 pounds, and tricycle landing gear, speed brakes, and flaps. Its maximum speed was better than Mach 1.2, at about 1,000 mph. In afterburner power, the rate of climb was over 30,000 feet per minute at sea level, with a service ceiling in excess of 50,000 feet.

The Talon's major safety factor was its two engines. If a pilot lost one, he still had the other to keep flying. It was possible to climb, fly acrobatics, and perform a complete mission on one engine. However, I was to fly on one engine only during practice. Tittle told me that if I lost an engine I was to immediately return to Edwards.

In the very unlikely event that I should have both engines flame out and couldn't get a restart, regulations told me to eject—bail out. The T-38 was too fast a plane for a pilot to attempt a dead-stick landing. I hoped I would never have an engine failure—the Talon was too beautiful a plane to abandon.

This would be my first flight in the T-38. Tittle would fly the backseat and evaluate my performance. I was well prepared and had memorized the Talon's normal and emergency procedures. The day came to fly the T-38.

"Any questions?" Tittle asked.

"No, sir," I replied without hesitation.

I was surprised by my inner calmness as we walked out to the flight line. I leaned forward against the weight on my backpack parachute and carried my helmet with oxygen mask under my arm. The calmness was as it should be. I'd flown many jet aircraft in my six years in the Air Force and had prepared myself well to fly the Talon. The training I had received from the test pilot school gave me the confidence that I could handle this new challenge. Colonel Yeager had told us at our introduction briefing that there was no substitute for experience in flying many different models of aircraft.

The crew chief smartly saluted as I approached the T-38. The portable air-start cart and the DC electrical power-input generator were humming in readiness. I read through the aircraft logbook and then started my walk-around inspection. Moving clockwise around the plane, I checked for oil leaks, dents, popped rivets—anything untoward.

Major Tittle attached his parachute straps, climbed into the rear cockpit, and started buckling in. I moved the entrance ladder forward and got into the front cockpit. Using the checklist I preflighted the cockpit:

Throttle: OFF
Landing gear: DOWN
Fuel boost: ON
Crossfeed: OFF
Battery and Generator: ON
Oxygen supply: CHECK
Warning and Indicator lights: TEST

I looked down at the crew chief, lifted two fingers, and rotated them overhead. Then I pressed the starter button and heard the turbine start to spin. At 12 percent rpm, I advanced the throttle to idle—the right engine ignited.

Soon the left engine was also running, and I signaled the crew chief to disconnect external power. He watched as I cycled the ailerons, stabilizer, rudder, speed brakes, and flaps. I removed the ejection seat and canopy safety pins, showed them to the crew chief, and he removed the chocks.

"Eddie Ground," I called over the UHF radio, "School 8-34 ready to taxi."

The first digit of every test pilot school call sign was the last digit of the aircraft type (i.e., T-33 was 3, T-38 was 8, F-101 was 1, F-104 was 4, F-106 was 6, and B-57 was 7). Every instructor and student was also given an individual number, which was used as the last two digits of their call sign. We all listened for School 4-01, Yeager's call sign when he flew the F-104 Starfighter, although it would be hard to miss his West Virginia drawl.

"School 8-34 taxi runway zero-four, altimeter two-niner-niner-two."

I adjusted the flaps to 60 percent, brought down the canopy, and locked it.

Approaching Runway 04 I called, "Eddie tower, School 8-34 ready to roll."

"School 8-34, cleared for takeoff, contact Eddie Approach airborne."

I lined up with the white stripe painted down the center of the 15,000-foot runway and braked to a stop.

"All set back there, sir?" I asked.

"Roger," Tittle responded.

"Maximum performance climb?" I asked.

"I'm with you, captain. You're a test pilot, feel it out!" he said.

I grinned beneath the oxygen mask and decided it was the last question I would ask. I advanced the throttles to full military power—100 percent rpm—and scanned the gauges for correct readings on the exhaust gas temperature, fuel flow, and oil pressure. The T-38 strained at the leash.

Releasing the brakes, I let the Talon roll a few feet and then punched in the afterburners. The acceleration pinned me to the back of the seat—faster, faster. At 140 knots I eased back on the stick and rotated the nose off the ground about 8 degrees. Around 160 knots the T-38's thin wings lifted her into the air and I quickly retracted the landing gear. As soon as the gear was in the well I retracted the flaps and accelerated to 500 knots. Over the middle of Rogers Dry Lake I pulled back on the stick, pointing the sharp nose into the sky at a 30-degree angle. The Mojave Desert dropped away rapidly. I streaked upward.

In just three minutes from brake release I was cruising at 35,000 feet and Mach 0.9.

I pulled both engines out of afterburner and spoke on the intercom, "I'd like to pull into some high-G turns."

Happy that I had worn an anti-G suit, I descended into a 6-G turn to the left and then to the right. Leveling off at 20,000 feet, I thought I would slow down and feel out the Talon's slow-speed landing characteristics.

"Let's go vertical," I heard Tittle say from the backseat.

"Well, here I go," I thought to myself, dreading the vertical maneuver.

I eased the nose over and did a clearing turn to the left and one to the right—other students might also be going vertical. The throttles were pushed to full military power and the speed was approaching 500 knots. I pulled the stick back, feeling the force of gravity building up, 3 Gs, then 4, and the nose was going up, passing 45 degrees. I checked the attitude indicator to make sure my wings were level. Up and up I flew. I tipped my head back and looked for a road below to confirm my alignment. The top of the artificial earth appeared in the attitude indicator, looking like the North Pole on a globe. The airspeed was slowing down—250, 200 knots—while the altimeter was circling clockwise in altitude so fast that I couldn't read the setting. Our test pilot school procedure called for me to take my hand off the stick. I looked upward at the dark blue Mojave sky and then the airspeed indicator. Speed: 150, 100, 50 knots—then we stopped in midair. Like rounding the top of a roller coaster, the nose dipped forward in a flash. I was hanging against my lap belt and shoulder harness and my feet floated off the rudder pedals. Within a second the Talon was aimed straight down at the Mojave and picking up speed like a falling rock.

"What a tremendous sensation!" I said on the intercom. "Wow, I liked that."

"Like riding a bucking bronco," Tittle chuckled. "Sometimes the T-38 falls over on its back, other times it peels off to the left or right. It depends on how accurately you can fly an exact 90-degree climb."

I flew several more vertical maneuvers—the thrill got better each time.

In the traffic pattern I found the Talon to be light to the touch and easy to fly. We made several touch-and-goes at Air Force Plant 42 in Palmdale, about twenty-five miles southwest of Edwards, followed by a simulated single-engine, full-stop landing at Edwards. The T-38 was simple to land, but the landing gear struts were so stiff that it was difficult to make a smooth landing.

Every time I flew the Talon, both as a student in the school and later as a fighter test pilot at Test Operations, I always saved some fuel to go vertical at the end of the flight. It was just too much fun.

All of our student instruction in flight testing was accomplished in the Edwards restricted area north of the base. These flights were all made in day VFR (visual flight rules) conditions. To keep our night and instrument currency, students scheduled weekend T-33 cross-country flights anywhere in the western United States. Classmate Jim Hurt and I decided to attend the 1964 National Air Races in Reno, Nevada, and see what lessons we could learn from civilians who were involved with air racing.

James W. "Jim" Hurt III, twenty-eight, was from Indianapolis, Indiana, and always had a big grin on his face; he was a happy-go-lucky type of guy. He had been a T-38 Talon flight instructor in Air Training Command at Laughlin Air Force Base, Texas, before attending the test pilot school. Hurt was married with two small daughters and lived on base just a block down the street from where Jan and I lived. During my year in the test pilot school Jim's wife frequently left home for days to places unknown, leaving him to take care of the children. It was a marriage nearing its end.

In 1964 Bill Stead, a Nevada rancher, hydroplane racer, and World War II pilot, came up with the idea of reviving the National Air Races to help celebrate the centennial of Nevada's statehood. The National Air Races previously had been held in Cleveland, Ohio, but were canceled after the 1949 event because of numerous crashes that resulted in fatalities on the ground.

Stead persuaded Reno businessmen to sponsor the races as part of a major air show that included the national aerobatics and balloon championships, a skydiving competition, and a performance by the Air Force Thunderbirds. Stead also talked ABC television into covering the races.

Competition was held at the Sky Ranch, a remote strip away from populated areas, where the runway was simply a 2,000-foot stretch of dirt. Pilots had wanted to take off from Reno Municipal Airport, fly to the course, and return to the airport after racing, but Stead had guaranteed ABC that takeoffs and landings could be filmed at the ranch. He threatened to disqualify any flyer that didn't use the makeshift landing strip, and the pilots reluctantly went along with it.

The races were staged at the Sky Ranch for the first two years. When Reno's Stead Air Force Base was closed in 1966, it was turned over to the city and renamed Stead Airfield; it has been the site of the Reno National Air Races ever since. Bill Stead was killed in a Formula 1 race in Florida shortly after the 1965 races. Ironically, Stead Air Force Base was named for his brother, Croston Stead, who had been killed in a crash while flying with the Nevada Air National Guard.

Daryl Greenamyer, twenty-nine, from Indio, California, was one of the pilots competing in the 1964 race. Greenamyer, a short, compact man with dimples and twinkling eyes, was in the test pilot class in front of me and had graduated a couple of months earlier. He was a civilian test pilot employed by

Lockheed who had flown the U-2 and would soon fly the SR-71 Blackbird. Greenamyer entered his F8F Bearcat in the Unlimited Class of races—the only restriction was that planes must be propeller-driven and powered by piston engines.

Greenamyer had significantly modified his Bearcat by installing a low canopy and sealed landing flaps. The flaps increased his safe landing speed about as much as the canopy increased his top speed. Visibility was so drastically reduced by the increased angle of attack and new canopy that Greenamyer felt it wasn't safe to land on the dirt strip at Sky Ranch. He chose the municipal airport for landing and takeoff, and thereby broke the rule that all airplanes must operate out of the race field.

Although Greenamyer broke a rule that caused him to be disqualified, he had made a smart decision from a test pilot point of view. In the school we were taught to set limits for ourselves. It was a good idea to set a personal do-not-exceed limit when venturing into the unknown. Test pilots had been killed at Edwards by accepting overly optimistic engineering estimates of aircraft stability and gone beyond their ability to handle the plane. Greenamyer had made a good call; he survived to race another day.

Over the next thirty-eight years, Greenamyer won nine championships at Reno. He broke the piston-engine speed record in 1969 and the low-altitude jet speed record in 1977 in an F-104 Starfighter made of surplus parts. Greenamyer once tried—unsuccessfully—to take wing from a frozen lake in northern Greenland in a World War II–era B-29 that hadn't been flown in forty-eight years. He was notorious for doing things in and with airplanes that ordinary people wouldn't even think of, and he has won the fastest division of the world's fastest motor sport more times than anyone.

Hurt and I found that air show performers visualized their performance many times before they went on the aerial stage. They took a few minutes before engine start to create a frame of mind that was relaxed and centered on the unique task at hand. Time was needed to focus on the upcoming sequence away from the thundering crowd. They closed their eyes, relaxed, and concentrated on the dangerous task ahead. Performers flew the entire display in their mind—every control input, every reference point and instrument reading, and visualized every maneuver right through final landing. When they were through they said their mind was clear and they felt relaxed and ready. This was also a technique that would work for test pilots. I've done it ever since then.

Immediately after each day's flying, all race pilots reported to debriefings within their respective group, which we sat in on. Honest discussions were held on the errors detected and methods to solve them. Test pilots had debriefed and kept track of "lessons learned" for years.

Race pilots have an adage that "any pilot, given the task of providing a display for the public, should set out to thrill the ignorant, impress the knowledgeable, and frighten no one." This neatly phrased saying was easy to state as an objective, but achieving that balance was far from easy. Ultimately, air show performers and test pilots must be prepared to fly only to their own self-set limitations. It's pleasant enough to hear compliments after a performance or to hear from flight test engineers when they look at your data, but it cannot compare to the feeling of having flown a truly professional, well-planned flight with absolute precision. Personal satisfaction is ultimately the best reward. Jim Hurt and I flew back to Edwards knowing we had learned a little more about our unique profession.

Chapter 7

Space Cadets

In July 1964 our class finished its training for the flight testing of aircraft. Everyone who had started the course six months earlier passed and was now qualified as a test pilot. The two foreign students returned to their countries, and Don Mallick, a NASA pilot, returned to the Flight Research Center located at Edwards.

Any pilot who was accepted in the Air Force school first had six months of aircraft test pilot training and then six months of space training and was humorously called a "space cadet." This six-month program included study of advanced mathematics, orbital mechanics, bioastronautics, aerothermodynamics, and spacecraft design. Besides checking out in the F-104 Starfighter, we would also get to fly two Century Series fighters (F-101 Voodoo and F-106 Delta Dart) that had been electronically modified as variable stability trainers. They were the first aircraft to have a computer on board to change the flying qualities of an aircraft. As space cadets we would experience weightlessness in a KC-135 tanker, high-G space reentry profiles in a centrifuge at Brooks Air Force Base, Texas, and a zoom flight to 80,000 feet in a standard F-104—heady stuff in the futuristic 1960s.

Our class was now made up of thirteen pilots. Five had a fighter background: Capt. David "Dave" Livingston and Capt. Richard "Dick" Strickland had flown the F-105 Thunderchief; Capt. Robert "Bob" Whelan, the F-102 Delta Dagger; Capt. Richard "Dick" Voehl, the F-100 Super Sabre; and I had flown the F-101B Voodoo. Three pilots had come from bomber/cargo squadrons: Capt. Ronald "Ron" Franzen had flown the KC-135 Stratotanker; Capt. Ernest "Ernie" Hasselbrink, the C-130 Hercules; and Capt. Gervasio "Jerry" Tonini, the B-47 Stratofortress. Capt. James "Jim" Hurt had been a flight instructor in the T-38 Talon, and Capt. David "Dave" Thomas had flown the H-21 Flying Banana helicopter. In addition, we had two Navy exchange pilots: Lt. John "Jack" Finley and Lt. Richard "Dick" Truly, who had both flown F-8 Crusaders. Finally, Fred Haise, a civilian NASA pilot, was already assigned to the NASA Flight Research Center at Edwards.

Our first real taste of space came in the form of a zero-G mission. A person can be displeased with gravity when they spill a glass of water, but just try pouring and drinking that glass of water without gravity. Likewise, after a long day on your feet a person can complain about their sore feet and blame gravity for it. But just try going for a walk without it.

As humans, we are born on earth and raised in a 1-G environment. Through the years we adapt to this force of gravity and become very comfortable with it. Since the first American entered space in 1961, the public has heard pilots and astronauts talk about zero-G. They realize that when a person is in orbit the normal pull of gravity is absent, a condition known as weightlessness. Both the astronaut and all his equipment will float around within the cabin if not secured in the space capsule.

To enter an environment with no gravity is so strange for the human body that astronauts must be trained for it. As space cadets we trained for the zero-gravity experience in a specially modified Air Force Boeing KC-135 aerial tanker.

The four-engine military KC-135 jet tanker used for zero-G flights was similar to the civilian Boeing 707 airliner except that most of the passenger seats were removed and replaced with heavy-duty padded cloth mats similar to those used in a school gymnasium. The forward section of the passenger cabin still had twenty regular airline seats for use during takeoff and landing. Separating these seats from the aft portion of the cabin was a cargo net to prevent us from bumping into the seats during zero-G maneuvers.

The area where we experienced zero-G was cylindrical in shape, approximately ten feet in diameter and forty-feet long, and completely surrounded with the gym mats, which covered all the windows. Wire screens covered several embedded lights that provided illumination during weightlessness and that were flashed by the pilot to warn us that he was either starting or ending a maneuver.

At 15,000 feet over the Mojave Desert, the KC-135 pilot accelerated to maximum speed in a shallow dive and then pulled the aircraft into a 45-degree climb angle. As the airspeed started to decrease, he pushed the control yoke forward and aimed for a zero-G indication on his cockpit accelerometer (G-meter), and the plane followed a parabolic path maintaining zero-G for thirty to forty seconds. As the aircraft approached 45-degrees nose low, he flashed the lights for a moment or so and then made a 2-G pullout to level flight.

Our instructors discussed the possibility of becoming sick during these unusual maneuvers, and so the airplane was appropriately nicknamed the "Vomit Comet." As jet pilots, everyone in our test pilot class had gone through military flight school and performed acrobatic maneuvers such as loops, aileron rolls, and stalls. In the test pilot school curriculum we had experienced the disorientation associated with spins, twisting, and turning in both pitch and roll as the aircraft plummeted toward the ground.

We had flown zero-G maneuvers in our training and fighter test aircraft, but we were restricted from more than a few seconds due to either aircraft limitations on the engine oil lubricating system or fuel transfer pumps. So at best we had flown for only a few seconds at zero G so as not to damage the aircraft. Also, we were always tightly strapped to our parachutes and ejection seat in a sitting position. In the KC-135 we were free to float around the entire cabin.

When it came to in-flight puking, our instructor said, "There are those who have and those who will."

He said it was nothing to be ashamed about and that it did not denote a character flaw. Medical data indicated that some people were simply more prone to motion sickness than others.

I remembered flying my first acrobatic maneuvers in Air Force flight school with an instructor in the backseat of the Beechcraft T-34 Mentor. After flying many loops and rolls, one time I felt queasy and placed the plane into straight-and-level flight. I took off my radio headset and removed the barf bag from a zippered pocket in my flight suit. I opened the bag and held it up to my mouth. Taking several deep breaths while fixing my eyes on the outside horizon, I gradually recaged my internal gyro and began to feel better. Slowly I recovered and up until now, six years later, had never experienced symptoms of airsickness again.

Two years before my flight in the KC-135, Jan and I vacationed in Hawaii. With an Air Force flight surgeon, a navigator, and both their wives, we hired a boat and crew to take us on a ten-hour deep-sea fishing trip in the Molokai Channel, which is southeast of Honolulu between the islands of Oahu and Molokai. The Air Force navigator was ill as soon as the ship left the dock; he spent the entire trip in a bunk bed below deck. Within an hour, his wife and the flight surgeon's wife also were sick and went below.

With my years of flying behind me, especially having flown acrobatics and in turbulence at night, I thought I would be immune to disorientation leading to seasickness. Unfortunately, the sea was very rough as we entered the Molokai Channel and prepared to fish for mahimahi. Eventually, after four hours, the rocking and rolling of the boat got to me and I lost it. Seeing sharks circling the boat at eye level due to swells didn't help either.

Sometimes people have a sympathetic response, whereby they tend to get sick after they see someone next to them vomiting. Witnessing her fighter pilot husband succumb to the curse of the sea, Jan followed my lead and deposited her breakfast in the Pacific Ocean.

The flight surgeon was the only one of the six of us to enjoy the fishing trip. He smoked his pipe, drank beer, ate our lunches as well as apple pie, and caught all the fish. The other five of us avoided him and took turns fighting for the four bunk beds below deck. Later we found that the flight surgeon had taken Dramamine before the boat left shore. The rest of us hadn't because we didn't think we'd get seasick.

So that our class of test pilots could determine our own sensitivities to zero-G flight, no drugs or pills were taken before our flight. Even though we had undergone a five-day pre-astronaut physical at Brooks, our flight surgeon said he could not predict with certainty who would get sick on this weightless flight. With that information in mind, I was quite apprehensive as we prepared for the first parabola.

Thankfully, the plan was to start easy and under as much control as possible. Therefore, for the first parabola, we were given a small Dixie cup full of water with which to experiment. As we went into weightlessness, the goal was to take a drink from the cup. Initially, my water and cup became weightless together and the water remained in the cup. As soon as I moved the cup toward my mouth, both the cup and water were placed in motion. Without thinking, I brought the cup to my lips and stopped. Unfortunately, the water continued in motion and sailed out of the cup all over my face. Everyone had a similar experience. The residual water, now in globs and spheres, meandered throughout the cabin getting everyone who collided with it soaking wet. This amusing adventure broke the ice, so to speak; we began having too much fun to even think about getting airsick.

On the second parabola, we were instructed to attempt to maneuver our bodies in zero-G flight. We tried swimming motions, swinging our arms and kicking our legs. If we drifted away from the padded wall beyond an arm's length, no amount of swimming motions would propel us through the sea of zero-G air in the cabin. Some pilots were upside down relative to the KC-135 floor, some were sideways, and others spinning like a top. All relationship between up and down was lost. I had to protect myself from getting kicked by other pilots who were also flailing around—we bounced off one another like billiard balls on a pool table.

For the third parabolic maneuver, we were divided into two teams. A three-foot-diameter leather-covered medicine ball weighing about a hundred pounds on earth was released in the middle of the cabin and floated free. Each team of pilots tried to push the ball toward the other team's position. Instead of the ball moving when I pushed it, I was propelled backward away from the action. It took team effort to wedge several pilots against the mat and gradually work the ball down the side of the cabin. I saw that teamwork was definitely needed in space flight.

For me the most exciting part of the zero-G demonstration flight was playing Superman. If I placed myself against the mat covering the tail section of the plane just as the KC-135 pilot started the zero-G maneuver, I could push both feet against the back wall like a person would do against the side of a swimming pool and zoom through the entire length of the padded area. I had had dreams in which I soared like a bird through the sky using only the arms and legs of my body to propel myself. In the KC-135 my dream came true, and for about forty seconds I was Superman.

On my own I discovered another interesting characteristic of gravity on the KC-135 flight. While the KC-135 pilot made a 2-G pull-up to a 45-degree climb, I tried to walk across the floor of the cabin. Now weighing about 340 pounds, twice my earth weight of 170 pounds, I felt like a gorilla walking through the jungle. I had to flex my knees to walk; it was too difficult to stand erect. As I lifted each foot it felt like a magnet was attached to my flight boot. When I relaxed my leg, my foot slammed down hard on the floor. I kept my hands out in front of me, fearing I would lose my balance and hurt myself falling down. It was much easier to crawl on both arms and legs, which distributed the extra weight better. I made a mental note to continue playing handball, stay physically fit, and keep from gaining excess weight in the future.

Fred W. Haise Jr., a thirty-one-year-old civilian NASA pilot, seemed to enjoy the KC-135 zero-G flight as much as we in the military did. He twisted, tumbled, and twirled like everyone else. I brought along my Argus C3 35mm camera on the flight and took several color slides of Haise and my other classmates floating in the cabin. All were having a ball. Like me, Haise found the zero-G experience to be thrilling and was sorry when the KC-135 returned to Edwards.

Haise was born in Biloxi, Mississippi. He had black hair, brown eyes, and a square jaw—the look of determination. Haise could not live in military base housing since he was a civilian, so he and his wife Mary and their three children lived in Lancaster, about thirty miles southwest of Edwards. Though a civilian now, Haise had been in the military before he joined NASA, completing his pilot training as a naval aviation cadet at Naval Air Station Pensacola, Florida.

After his flight training, Haise served as a tactics and all-weather flight instructor in the Navy's Advanced Training Command at Naval Air Station Kingsville, Texas. From 1954 to 1956 he was a Marine Corps fighter pilot at Marine Corps Air Station Cherry Point, North Carolina. Then from

1957 to 1959 he was a pilot with the 185th Fighter Interceptor Squadron in the Oklahoma Air National Guard, while a student at the University of Oklahoma.

When he completed his degree, Haise worked for NASA for three years as a research pilot at the NASA Lewis Research Center in Cleveland. During that period, he also worked for the Ohio Air National Guard and for the Air Force as tactical fighter pilot and chief of the 164th Standardization-Evaluation Flight of the 164th Tactical Fighter Squadron at Mansfield, Ohio. Since Haise had flown in three of the military services, my classmates kidded him that he couldn't make the grade in any of them. Nothing was farther from the truth; Fred Haise was an excellent pilot. We all knew he would go far in his flying career.

In late fall our class visited Brooks Air Force Base for discussions with Air Force aeromedical personnel on the physical aspects of space flight. The highlight of the trip was several rides in a high-G centrifuge. Now we would encounter more Gs than you could ever pull in an aircraft. We'd be subjected to Gs inclined on a couch so that the force encountered would be from chest to back rather than head to toe as in an aircraft. We squeezed a handle in our right hand that controlled a set of lights arranged in a cross. While we were pulling Gs the operator of the centrifuge would illuminate one light on the vertical row of lights and one on the horizontal. Our task was to move the control handle in such a way as to move both lights to the center, confirming that we were still conscious. In addition, there was a "dead man's" switch in the handle. If a pilot released the handle the centrifuge would automatically stop. A doctor watched a TV monitor pointed directly at the pilot's face, and if the subject was unresponsive he could also halt the run.

For my first run I was subjected to 8 Gs for about five minutes. I had pulled that many Gs in an aircraft before, but never for that long, and at the end started to get tired. The next run was 10 Gs for two minutes followed by 12 Gs for thirty seconds. I noticed that the extreme force on me caused my tongue to fall back in my throat and I had a hard time talking. Also the skin on my upper cheek was pulled back so hard that I couldn't focus my eyes. Overall, it was very difficult to think and react under heavy-G forces.

Finally, on the last run I went up to 15 Gs and immediately returned back to rest. I was relieved when the experiment was over. After getting out of the capsule I had a difficult time walking, as my sense of balance was lost. I got dry heaves and gagged for several minutes. Each of us was given a certificate upon completion. Called the Order of the Elephant, it showed an elephant standing on a pilot's chest. And that's exactly what it felt like.

Flying officers from the French air force came to visit the test pilot school about a month before our graduation. Colonel Yeager showed them the school aircraft, our facility, and invited them to a dining-in at the Officers Club. A dining-in was a formal dinner in which officers wore a mess dress uniform, the military equivalent to a civilian tuxedo. A gorgeous meal was prepared, wine flowed freely, and toasts were made to the military chain of command. After dinner, cigars were handed out and a guest speaker tried to keep the unruly crowd entertained. Sitting at the head table next to Chuck Yeager was none other than the illustrious "Pancho" Barnes.

The saga of the Air Force Flight Test Center would not be complete without mention of one of its most enduring friends, the former Ms. Florence Lowe, known to all the world by her favored nickname, Pancho.[1] Never officially a part of the Edwards community, or ever directly connected with the Air Force, she nevertheless spent many years as one of its most enduring champions and unswerving friends. She has become familiar to the general public as the colorful, swashbuckling friend of America's best-known test pilots. But the aviation community had always known her as a skilled professional and one of the respected figures in the early golden age of flight. Long before Pancho Barnes ever set foot in the Mojave Desert, she had already made her own mark in the progress of American aviation and women's role within it.

At the age of eighteen, Florence Lowe wed the Reverend C. Rankin Barnes, a prominent Episcopal minister, and settled down to the duties expected of a proper clergyman's wife. In due course their son, William, was born. Not long afterward, however, the young bride's self-reliant personality asserted itself in dramatic fashion. Abandoning church and child in the early 1920s, she disguised herself as a man and boarded a freighter headed for Mexico, where she signed on as a banana boat crew member. Once the ship was safely docked at San Blas with a cargo of bananas and contraband guns, she jumped ship with a renegade sailor and spent four months roaming through the revolution-torn interior. Somewhere along this trek, while riding a donkey, her comrade nicknamed her "Pancho" for her fancied resemblance to Don Quixote's faithful companion, Sancho Panza. She was delighted with her new nickname and kept it for the rest of her life.

After Pancho returned to California from her adventures in Mexico, she was restless and became bored easily with the routine of normal living. When a cousin invited her to go with him to an airfield where he was taking flying lessons, she fell in love with flying and quickly got her pilot's license.[2] The young aviatrix burst onto the national aviation scene barely a year after her first solo flight. In August 1929, she joined nineteen other women in the Women's Air Derby, a transcontinental air race from Santa Monica to Cleveland. This was the first Powder Puff Derby, which is still being flown today. She got as far as Pecos, Texas, before she ran afoul of the casual airfield-management practices of the day, colliding with a truck driving down the runway. Pancho was unhurt, but her broken airplane put her out of the race for that year.

By then her growing reputation enabled her to sign on with Union Oil Company for a three-year stint of demonstration flights and promotional work in return for sponsorship in many of the air races of the day. She returned to the Powder Puff Derby the following year in a powerful Travel Air Model R "Mystery Ship"; it was a low-winged speedster with huge wheel spats that has been called the most beautiful of the great racing airplanes. Blasting across the route at an average speed of 196 mph, she took the world's speed record for women away from Amelia Earhart.

All good times come to an end, however, and so it was for Pancho's dizzying world of flying and glamour. The nation settled ever deeper into the Great Depression, and the fortune that Pancho inherited from her mother began to melt away, hastened by an indecorous conflict with her husband in San Marino. Still officially married to the hapless churchman, she traded most of her surviving assets in 1935 for a small quarter-section ranch in the desolate reaches of the western Mojave Desert. Her test pilot work for Lockheed had brought her up to the high desert of the Antelope Valley where she spotted a small green jewel of an alfalfa farm located between Rosamond and Rogers dry lakes. One of her first actions was to carve out a rough landing field so she could continue flying her plane and so her pilot friends could come up to visit. The ranch was a working one, and her ranch hands were soon not only raising alfalfa, but horses, pigs, and dairy cows.[3] Pancho Barnes took Bill Barnes, her twelve-year-old son, and settled down to the unlikely life of a rancher in the high desert.

Soon Pancho began to expand her operations, enlarging her herd of milk cows and selling dairy products throughout the Antelope Valley. The remains of her family money went into ranch improvements, and within a few years the ranch had grown from 80 to 368 acres. She enlarged the ranch house and built a swimming pool—an exotic touch for the late 1930s. As war clouds gathered abroad and the nation began to shake off its peacetime torpor, the Air Corps began a long-overdue expansion. The nearby bombing range grew larger, the government bought up great amounts of land, permanent buildings went up, and officers and enlisted men began to appear in larger numbers.

When World War II arrived in the high desert, Pancho was swept along with the current. The gunnery range became Muroc Army Air Field, a huge expansion began on the western shore of the lake, and permanent runways were built for year-round use. Suddenly a major military installation lay only three miles down the road. Pancho was delighted at the new turn of events. Patriotically, she made her ranch available to off-duty fliers. Officers—and especially pilots—were welcome in her swimming pool; often they stayed for dinner and discussions about flying went on far into the night. Pancho offered her horses for the recreation of those who could ride, and bought more. By degrees, the desert exile became a hostess.

In retrospect, it all had a kind of inevitability about it. The airmen loved Pancho's party atmosphere, while the opportunities for other recreation were severely limited. Wartime money was suddenly available, visitors always needed a place to stay, and Pancho had plenty of room to expand. A bar and restaurant appeared, then a dance hall, another bar, and a coffee shop. Most of the booze came up from Mexico in Pancho's plane and was dispensed freely; the more expensive stuff stayed under lock and key. The airstrip was enlarged and lighted for the increasing number of guests and friends who flew in, and a motel was built for their convenience. Soon Pancho found herself the proud mistress of the Rancho Oro Verde Fly-Inn Dude Ranch.

Ever more boisterous, profane, and swashbuckling, Pancho proceeded to have the time of her life. Almost gleefully, she allowed time and the dry desert air to transform her youthful appearance into the storied homeliness by which most remember her. To compensate, Pancho imported an ever-changing bevy of attractive hostesses to serve the weary airmen. Even the name of the ranch reflected the wartime gaiety, soon being nicknamed the Happy Bottom Riding Club in salute to the growing number of skilled and satisfied riders. Pilots were always her special comrades, and in the natural course of events a stellar array of high-ranking officers appeared at the ranch and soon became her friends.

Thus it was natural that when peacetime came and Muroc (soon to become Edwards Air Force Base) became the center for the nation's leading experimental flight testing center test pilots would replace the wartime fliers and the party would go on. Pancho's place remained popular for the same reasons it always had—in an isolated area of limited resources, men with heavy responsibilities needed a congenial place to relax. Although many stories about Pancho and her hostesses are told with a knowing wink, it is also true that off-duty pilots love to do one thing above all—talk about flying. And there was plenty of that at Rancho Oro Verde.

Pancho was a staunch friend and confidante to many of the young professional fliers of the day—Chuck Yeager, Pete Everest, Jack Ridley, and many others. Friends with those that she liked, that is. Those whom she did not, or who carelessly patronized her, were swiftly and profanely shown the door.

With Chuck Yeager, a bond was formed that lasted her lifetime. Recent books and movies have glamorized the friendship between the sonic-busting test pilot and the high-flying hostess, but in truth it began earlier when Pancho found out that the young captain was also an avid outdoorsman. Several hunting and fishing expeditions sealed their friendship before Captain Yeager had been chosen to bring the X-1 supersonic program to its ultimate success. When he did so, on October 14, 1947, Pancho was one of the few who knew about the official secret. Yeager won a free steak dinner for that feat, thereby starting a tradition for all pilots celebrating their first supersonic flight.

Soon the entire atmosphere began to change, however. The reasons were many: conflicting requirements, personality clashes, and some genuine misunderstandings. The immediate catalyst was conflict over airspace, which was becoming increasingly crowded with large numbers of new aircraft being tested and the private airplanes of Pancho's guests. The borders of the base were already pressing hard on Rancho Oro Verde, and a master plan had already been written calling for it to expand to its present western boundary. Sooner or later, something would have to give. But the times were changing as well. The Happy Bottom Riding Club was doomed in any event.

It was not long before condemnation proceedings were filed against Pancho's property based on the fact that the ranch lay in a direct line with a proposed extension of the test center's main runway. There were genuine air-safety considerations as well. But the situation was greatly worsened by a complete lack of rapport between the principals, and conflicts soon escalated into name calling, unjust accusations, and ultimately into a flurry of acrimonious lawsuits. In the middle of the fray, coming at the worst possible time, a nighttime fire of unknown origin destroyed the ranch complex.

Pancho eventually won a considerable sum in the courts. She established herself on a new spread in another remote area, vowing to rebuild and continue as before. But much of the settlement went into bad investments and gifts to friends. She had a big heart and gave money away. Pancho had lost not only her ranch and livelihood but also a lifetime's accumulation of irreplaceable souvenirs and valuables. Perhaps worst of all, though, was the rift with her beloved Air Force. But she still had Chuck Yeager, and he had invited her to our dining-in.

The dining-in went on without a hitch even though Yeager toasted the king of France. France had not had a king for more than 175 years, so the Air Force officers got a good laugh at his remarks. Only polite smiles were seen on the faces of the French officers.

During the toasts I watched Pancho Barnes. She was sixty-three years old and the only woman in the entire crowd. Female Air Force officers were rare in 1964. She laughed as Yeager spoke and seemed to be enjoying the personal attention. The years of hard living and exposure to the sun had not been good to Pancho. She had a huge round head, the largest I had ever seen on a woman. Her face was wrinkled and red— she was not very attractive. Nevertheless Pancho was a unique figure to see in person and she played a large part in the early history of Edwards Air Force Base.

As a student in the space school, my zoom flight would be the high point of the twelve-month course and my last flight as a student in the school. I'd fly the F-104 in a pressure suit to an altitude where few pilots had been privileged to fly—the rarefied reaches of the atmosphere above 80,000 feet. The purpose of my flight was to acquaint me with the experience of space flight.

As I sat in the cockpit of the F-104 waiting for the Edwards control tower to give me permission to taxi onto the runway, I thought about the aviation history made at Edwards by such great test pilots as Tony LeVier, Bob Hoover, and Scott Crossfield. I also remembered in my youth reading about historic airplanes from the first jet Bell P-59, the rocket powered X-1, and the record-setting X-15, which had all first flown here at Edwards. From my position on the taxiway I saw the majestic XB-70 Valkyrie parked on the ramp and the hangar where the YF-12 Blackbirds were housed. I felt awed by the aviation history made on this desert base and pleased to have the opportunity to attend the test pilot school.

The full-pressure suit, in which I'd spent the last hour and a half, was causing me to sweat profusely. The view from the pressure suit helmet was like looking at the world from inside a fishbowl. The white plastic helmet and faceplate were almost the same size as the small F-104 canopy that I had just closed. I could only move my head a few inches from side to side before it thumped against the canopy.

As I sat cooking in the canopy that was acting like a greenhouse under the intense Mojave Desert sun, I felt confident in my ability as a pilot. I'd logged thousands of hours in Air Force fighters from the F-86 to the F-101B, but I'd never flown an F-104 up to 80,000 feet. Would my experience count for much up in the sky in a world that belonged exclusively to the X-15 and NASA space capsules? I'd flown the F-104 many times in the previous months to get the feel of the flight controls. For many hours I'd thought through the profile that I'd fly. Like a race pilot I'd gone over in my mind all foreseeable emergencies I might encounter. I'd practiced dead-stick landings in case the J79 engine wouldn't restart after the reentry portion of the zoom. However, there is always a little bit of doubt when you're attempting something significant that you've never done before. When the chips were down, would I hack it?

I remembered when I first started flying functional check flights in the F-101B Voodoo. There was a feeling in the pit of my stomach that said, "This is it. You are flying the plane to the very edge of the F-101B's maneuvering boundary . . . don't lose it!"

If I pulled too much on the F-101B's control stick, I'd pitch up, stalling the wings and falling into a spin. If I didn't pull hard enough, I wouldn't be able to set the electronics in a manner so that squadron pilots could get the maximum capability out of the Voodoo. It was not fear I felt but a mixture of apprehension about and anticipation of the unknown. That day the feeling was the same.

If I overcorrected at the top of the F-104 zoom, I'd be uncontrollable in seconds. This had happened to one of the students in the class in front of me. Lt. Patrick "Pat" Henry, a Navy pilot, lost control at the top of the zoom, fell in a spin, and eventually ejected from his Starfighter. My fate would be the same, if I were not precise in my planning and control of the plane. If the engine failed to restart coming down the backside, I'd be committed to a real flameout pattern. There were many ifs—many things that could go wrong. I knew all the emergency procedures and was confident. But, when it's for real and there's no room for error and no second chance, would I make the right decision?

"Zoom 5, you're cleared onto Runway 04 to hold."

The tower's call interrupted my thoughts. I glanced at my checklist.

Canopy: LOCKED
Warning lights: OUT
Speed brake: IN

The salty sweat dripped into my eyes, but I'd be cool up there in the wild blue yonder where I'd be going.

Flaps and trim: TAKEOFF POSITION
Attitude indicator: CHECK

A quick glance to my left confirmed that my chase aircraft, another F-104 with a call sign of Zoom Chase, was in position and ready for takeoff. He'd chase me until the pull-up point, then rejoin in formation as I descended through about 20,000 feet in order to chase me through the traffic pattern. He'd observe my exterior appearance, be ready to offer any assistance I might need, and help keep me clear of other traffic, since most of my attention would be focused inside the cockpit cross-checking instrument readings.

The J79 gave out its characteristic howl and roar as I eased the throttle full forward and back again to idle. It's a unique sound that is unmistakable to any military pilot and guaranteed to turn heads. It was a sound of power—and a sound of death due to the many accidents in the Starfighter.

I'd looked forward to flying the F-104 ever since I first saw one on the ramp at Webb Air Force Base as a student pilot. The plane looked as if it were doing Mach 2 just sitting on the ramp.

"Zoom 5, winds are calm, you're cleared for takeoff," called the Edwards control tower operator.

No time for other thoughts now—time for action. I'd be very busy during the next forty minutes.

Throttle: FULL FORWARD
Flight controls: CHECKED
Throttle: MIMIMUM AFTERBURNER
Brakes: RELEASED

I got a good afterburner light, and then I selected maximum afterburner. The traverse Gs pressed me against the parachute strapped to my back. What a mighty plane! Flight control stick aft at 100 knots, nose wheel raised at 150, airborne at 175. Landing gear up quick before I reached 250 knots and ripped the gear doors off—then flaps up. Passing 400 knots, I raised the nose slightly to start my climb and brought the throttle back around the horn and out of afterburner. I started a gentle right turn to the east and climbed at 450 knots, waiting for the Mach to build to 0.85.

The chase pilot confirmed by a thumbs-up that my Starfighter looked fit to climb and I continued my mission. Climbing toward the morning sun, which glistened off the highly polished needle nose, I had a few seconds to enjoy flying in this beautiful aircraft. As a boy I remembered being in love with Lockheed aircraft such as the P-38 Lightning. Later I would fly the Lockheed T-33 in flight school. But the Lockheed F-104 Starfighter was the ultimate for the moment. In my estimation, Lockheed built the best and fastest aircraft in the world. Speed was king. Now the triple-sonic Lockheed YF-12 and SR-71 Blackbirds were the planes to beat.

No time to daydream, I must keep my mind on this test mission. Climbing at 0.85 Mach, I leveled off at 20,000 feet passing abeam of Three Sisters Dry Lake. It was time to dump cockpit pressurization and inflate my pressure suit. Since I would be shutting down the J79 engine during the zoom, all pressurization would be lost. If my pressure suit failed at 20,000 feet, I would still have time to repressurize the cockpit, abort the mission, and return to Edwards. Slowly the suit inflated, but I felt like a fat man in a telephone booth, unable to move much in any direction.

As I continued to climb to 35,000 feet, I saw Baker Dry Lake in front of me. The lake was about one hundred nautical miles from Edwards and my turning point for the run back in the supersonic corridor. I made a gradual 180-degree turn to the left, glancing over my shoulder to confirm that my chase was still hanging in there.

Rolling out on a heading of 275 degrees, I pointed the nose toward the town of Tehachapi. Moving the throttle again around the horn, I selected maximum afterburner and eased the control stick forward ever so slightly to unload the 1 G of level flight, helping the Starfighter blast through the transonic zone. The Mach meter slipped through Mach 1.0 with no physical sensation in the plane whatsoever. The Mach was really climbing fast now, 1.3 then 1.4.

I glanced up at the clear blue sky above the canopy bow. If all went well, in a minute or so I would be pulling a contrail that went up at almost a 45-degree angle into the gorgeous blue Mojave sky. The contrail would be visible for hundreds of miles.

Trying to push the throttle harder against the forward stop, I hoped to get every last ounce of power from the engine as I accelerated to Mach 1.7, then 1.8. The F-104 was in its best operating area now and the Mach was climbing fast. At 675 knots indicated airspeed, I started a gradual climb to 38,000 feet. It was a tremendous feeling to be going faster and faster. The chase aircraft was miles behind me now as I approached Mach 2.15. I let the Starfighter accelerate as long as I dared, since I wanted every bit of energy I could get. The more speed I built up, the more altitude I would get over the top.

It was time for one last glance at the checklist. I had penciled a note at this point: check gloves. Capt. Jerry Tonini, my classmate, had one of his leather gloves start to balloon in the thumb just before he started his pull-up. Fortunately, he caught the problem in time. Had the glove broken open, he would have lost all suit pressure. Then he would have become unconscious in only four to five seconds and probably crashed.

The compressor inlet temperature approached 155 degrees Celsius, the maximum allowed. One last check on the fuel: 1,200 pounds was the minimum I must have before starting the zoom and still be able to recover with a safe reserve to land at Edwards. The needle pointed to just less than 1,200 pounds remaining.

Go for it, I thought, pull up. At that moment the image of Colonel Yeager wrapped in all of his bandages flashed before my eyes!

I began back-stick pressure at a rate of 1 G per second. When the G meter reached 3.5, I maintained the pressure constant and transferred my attention to the attitude indicator in the center of the instrument panel. As I reached 40 degrees of pitch attitude, I slowly eased off the back-stick pressure and stopped the nose rise at the preplanned angle of 45 degrees. I monitored the J79 exhaust gas temperature (EGT)—didn't want to "overtemp" the engine.

Quickly I glanced at the altimeter—the needles were spinning so fast that they blurred. I'd passed 60,000 feet, the EGT was 615 degrees—maximum allowed. I began to retard the throttle to hold EGT constant. I brought the throttle into cutoff while passing 67,000 feet.

The angle of attack approached 8 degrees, so I pushed forward on the stick and felt weightlessness coming on. Even though my shoulder harness was firmly tight and locked on the ground, I felt my pressure suit lift off the ejection seat and my helmet touch the canopy.

Just approaching the peak, I treated myself to a panoramic view of the earth. Most of the flight so far had been head in the cockpit—fly the gauges. The sky was very dark blue—almost black. I could see the Pacific Ocean coastline in front of me, although still over a hundred miles away. There was the smog of Los Angeles down at the left over the mountains and at my right the Bay Area where my old fighter squadron was located. Sightseeing was over—I'd topped out at "Angels 80," about 80,000 feet. It was so quiet in the cockpit I thought I heard my heart beat.

Speed brakes out, airspeed started to build up fast as the light brown Mojave Desert came back into view. I was now diving straight down with Rogers Dry Lake directly off to the left. Passing 25,000 feet, I remembered I needed to restart the J79 engine.

Start Switch: START
Throttle: IDLE
Fuel Flow: CHECK
EGT: WATCH FOR RISE

The engine EGT started to rise, I had a good light. With the engine running, I started a turn back to the Edwards runway when I noticed a silver flash on my faceplate. It startled me until I realized it was a drop of sweat from my forehead.

I passed high key at 20,000 feet directly above Edwards Runway 04, where I had started this flight about a half hour earlier. I'd be landing out of the same pattern that was used by the X-15, 300 knots airspeed and a 20-degree dive angle. With just 90 degrees to turn to line up with the runway, I was at

15,000 feet. I had flown the pattern many times before and felt quite comfortable. Rolling out with the runway centered at 6,000 feet, I had the 15,000-foot landing strip centered in front of me. I started the stick coming back for the flare and lowered the landing gear at 250 knots. Then I checked the gear down and locked it just before touchdown.

The tires squealed as they burned rubber on the painted white line that crossed the runway at the 10,000-foot remaining marker. As I lowered the nose gently onto the runway and pulled the drag chute handle, my chase pilot flashed past on the right side of the canopy as he made a low approach.

More sweat dripped into my eyes as I looked up at the contrail my zoom had etched against the blue Mojave Desert sky. I had safely returned from the edge of space.

Chapter 8

Flight Test Operations

After graduation from the test pilot school, two of my classmates, Dave Livingston and Bob Whelan, and I were transferred to the Fighter Branch of Flight Test Operations, which was located just down the Edwards ramp from the test pilot school. Yes, I was interested in traveling in space, even to the moon, the fanciest whistle I could imagine. Like others I applied to the NASA space program after I graduated. But in 1966 there were about four hundred applicants and only around fifteen were selected. It felt like playing the lottery. Though I was disappointed at first on not getting in, I later realized that flying aircraft was my principal calling in life. When my test pilot colleagues Bill Anders, Fred Haise, and Joe Engle were accepted and traveled in space I felt like a vicarious participant.

I considered the Test Operations assignment the best job you could get as a junior captain in the Air Force. But the certificate of completion from test pilot school didn't hold much weight at Test Ops, since every pilot had graduated from one of the military testing schools. Livingston, Whelan, and I were now the "new guys," bottom of the heap, chasing the whistle again. We inherited one of the most undesirable jobs—barrier testing.

The barrier was a device used by the Air Force to halt aircraft on the runway in a short distance in case of an aircraft emergency or inclement weather. It was equivalent to Navy arrestments on board a carrier using a tailhook. Whereas the Navy equipment was designed to stop an aircraft on the flight deck in only a couple hundred feet, the Air Force landing fields have runway space of a thousand feet or more. Air Force Century Series jet fighters were not initially designed with a tailhook, but some were retrofitted later.

Barrier testing was done on the ground. The test pilot could not log flying time nor could flight pay be obtained. None of the old guys did barrier testing—they had moved on to important flying assignments.

All the barrier testing was done at Edwards South Base, a facility about two miles south of the main base and constructed in the 1940s. It consisted of a short 8,000-foot runway and several small hangars. The base was used as the primary test facility until the main base was built in the mid-1950s.

In case of a mishap, an Air Force fire truck and ambulance were stationed near the runway and on guard because of the inherent dangers involved with barrier testing.

The fastest route to South Base from Test Operations was to drive directly across the busy 15,000-foot main runway. Personal autos were not allowed on the taxiways or runways. Therefore the standard procedure was for an ambulance driver to pick up the barrier test pilot at Test Operations and, with the use of the UHF radio in the ambulance, get clearance from the control tower to cross the main runway, Runway 04/22, at midfield and drive to South Base. The test engineer was already at South Base preparing the barrier for the test.

As test pilots we realized that at some time in our career we might have an in-flight emergency that could cause us to crash at Edwards. We would then be transported to the hospital in an ambulance. Barrier engagement was the only type of testing at Edwards where you went *out* to the test aircraft in an ambulance.

Arriving at South Base, the new guy would find the most dilapidated jets he had ever seen. The aircraft were all relics of previous flight testing many years earlier and were not meant to ever fly again. Some of the aircraft had no canopies, and the F-101 even had wood nailed to the side of the fuselage. The cockpits were in shambles with cobwebs and the remains of birds' nests. The faded instrument dials and switches were practically unreadable.

The aircraft scheduled for me to test that day was an F-100 Super Sabre. An Air Force mechanic stood on the entrance ladder, which was attached to the side of the cockpit, and helped me strap in and start a very sick and groaning engine.

"Don't let the airspeed get up to flying speed," the mechanic warned. "The flight controls haven't been operational for years."

The test engineer consulted his test plan. "Hit the barrier at exactly 113 knots," he screamed to me over the whine of the pulsating engine, "fifteen feet right of runway centerline."

As I advanced engine power at the end of the runway, I aimed the F-100 at a small cable strung across the runway about 3,000 feet away. Dust swirled in the open cockpit. I connected my oxygen mask even though the aircraft had no oxygen, and then pulled my helmet visor down to keep the dust out of my eyes.

I released the brakes and went charging down the runway. Speed built up very quickly, as the aircraft was light with only a partial fuel load. As I approached the 113-knot engagement speed, I pulled the throttles back toward idle and dropped the tailhook. It was purely a guess where to aim so as to be fifteen feet right of centerline. As I crossed over the metal cable strung across the runway, I felt a thump-thump as the nose wheel and then the main landing gear crossed the cable. I kept my feet off the brakes and just used rudder movement to keep the aircraft headed straight.

I waited for a feeling of deceleration, which would mean the tailhook had grabbed the cable. I waited longer. Soon the end of the runway came up and I was still speeding along at just under 100 knots. Fortunately Rogers Dry Lake started at the end of the runway and I had miles and miles to roll out. The aircraft bounced and shook like an automobile crossing railroad tracks as I left the runway and went screaming over the lake bed. It gradually slowed down, and I used nose-wheel steering to make a wide 180-degree turn and taxi back to the ramp.

The test engineer explained that the tailhook had bounced up and down after deployment and skipped over the cable, hence no engagement.

"Glad you didn't use brakes as you crossed over the end of the runway" he explained. "Some aircraft landing gear will collapse if you use hard braking while the plane rolls off the runway and onto the lake bed."

I wish I had known that before I made the test run, I thought to myself.

"We'll reschedule another run tomorrow," he added.

As I sat in back of the ambulance on the return trip back to Test Ops, I looked forward to the next class of the test pilot school to graduate. Then an even newer "new guy" assigned to fighter test would take his turn at barrier testing.

Two years later, Art Peterson, one of the experienced civilian Lockheed SR-71 test pilots, conducted an antiskid braking evaluation on the main runway at Edwards. Peterson, flying the number one prototype, serial number 64-17950, accelerated to high speed and attempted to use maximum braking to determine the Blackbird's minimum stopping distance on a wet runway. Unable to stop on the existing runway, he continued to use heavy braking as the SR-71 went off the runway and on to Rogers Dry Lake. In this case, the tires blew out and the landing gear collapsed. The aircraft caught fire and was destroyed. A multimillion dollar aircraft was lost (although Peterson survived and flew U-2s for a number of years). I guess that "old guy" had never paid his dues to learn the fine art of barrier testing like we "new guys."

Although the T-38 was a dream to fly, I looked forward to a chance to fly the fighter version called the F-5A Freedom Fighter. Most of the initial flight testing of the F-5 had already been accomplished when I arrived at Test Operations, but several follow-on programs remained. Some test pilots considered the Freedom Fighter a tough little scrapper, while others called it a pint-sized Bantam jet.

The first overseas order for F-5s was from Norway, for sixty-eight aircraft. My classmate, Capt. Robert E. "Bob" Whelan, thirty-three, now with me at Test Operations, headed up the evaluation of their aircraft. The Norway Freedom Fighter differed from standard F-5s by having a heated windshield, a tailhook, and provision for jet-assisted takeoff (JATO).

The F-5A was optimized for an air-to-ground role and had a very limited air-to-air capability. As an afterthought, two Colt-Browning internal 20mm cannon were installed in the nose, and two AIM-9 Sidewinders could be carried on the wingtips.

Quite a few other upgrades were made to change the initial T-38 trainer design into a fighter. Two General Electric J85-13 turbojets were installed, boosting the top speed to Mach 1.4. Hard points were added to both the fuselage and four underwing stations so that up to 6,200 pounds of ordnance or fuel could be carried. A 50-gallon fuel tip tank could be carried at each wingtip, and a 2,000-pound bomb or 150-gallon fuel tank could be carried on the fuselage centerline pylon. Underwing loads on the four wing pylons could include two 150-gallon fuel tanks and four air-to-air missiles, Bullpup air-to-surface missiles, bombs, or rockets.

Maj. Michael J. "Mike" Adams, thirty-five, one of the Fighter Branch test pilots who had tested the F-5A and would later go on to fly the X-15 rocketship, briefed me on the flight characteristics of the single–seated Freedom Fighter. I read the flight manual, and Adams quizzed me on the emergency procedures.

Right away I noticed some major changes in the F-5 compared to the T-38. Vision over the long nose was excellent except that I sat so far back (equivalent to sitting in the backseat of the T-38 next to the engine air intakes) that I felt like I was driving from the rear seat of a Cadillac. The instrument arrangement and cockpit roominess were outstanding, one of the best I had ever seen in an airplane.

I cranked up the two General Electric J85 turbojets and taxied to the end of the Edwards runway. The distance from the ramp to the taxiway alone was about a mile, and from there to the runway was another two miles. Just as I turned onto the runway the red "tip tank empty" lights started to flash, for I had used most of the tip-tank fuel just getting to the end of the runway. F-5A pilots called the jet short-legged, meaning that it had a small operating range without in-flight refueling. Because the maximum gross weight for flight was slightly over 20,000 pounds (8,000 pounds greater than the T-38), it took a lot of power to keep the F-5 flying at combat speed. Hence, a great amount of fuel poured through the GE engines, and the Freedom Fighter couldn't get too far from home or a tanker. This was a characteristic that limited its effectiveness in Vietnam.

The stick force required to rotate the nose for takeoff was very great, because the aircraft was being flown with a much more forward center of gravity (due to the gun in the nose) than the T-38. Also, the control stick was placed further aft, closer to the ejection seat, making it difficult to get full aft travel.

In the air, however, the F-5A was an absolute delight. Visibility through the windshield and canopy was impressive, and flight controls were light and responsive. The Freedom Fighter had very docile handling qualities—it turned like a charm with just light buffet preceding wing stall. And, best of all, it could go vertical like the Talon.

After some aerobatics I returned to the traffic pattern for a couple of touch-and-goes. Coming down final I again noticed the long nose out in front of me, but I still had a clear view of the runway. The long nose looked strange, and again the control stick was just a little too far aft to feel comfortable. The T-38 trainer had stiff struts and high-pressure tires—hard to make a smooth landing with that combination. The F-5A, on the other hand, had large fat tires and soft struts, so I greased the sleek machine onto the runway. The extra smooth landing was kind of a waste, though: I was in a single-seated aircraft and, unlike the T-38, didn't have a pilot in the backseat to impress.

Mike Adams had warned me about the Freedom Fighter's drag chute. The chute was so effective, he said, that if I brought the airplane to a complete stop on landing, I wouldn't have enough power to taxi off the runway unless I used afterburner. I deployed the chute just after touchdown, waiting to experience this unusual feature. Adams was absolutely correct—the plane immediately slowed down. As the airplane nearly came to a stop, I angled over to the right side of the runway and then, just before stopping, made a sharp 30-degree turn back to the left. Just as Adams recommended, I released the drag chute and watched it soar well off to the side of the runway. Overall, the F-5A was a super machine, a pilot's plane.

Although all early F-5A production was intended for the Military Assistance Program (MAP)—which sold planes to small countries that didn't have aircraft production facilities and couldn't afford expensive fighters—the Air Force requested two hundred F-5s for use in Vietnam. The Air Force, which had perceived no need for a lightweight fighter, had experienced heavier than expected attrition in Southeast Asia, and the F-5 promised to be available with a relatively short lead time. The single-seater was to have been designated the F-5C, the two-seater F-5D. The Defense Department initially turned down the request, but an Air Force request for combat evaluation in Vietnam was approved.

The flight evaluation was given the code name Skoshi Tiger, after how *sukoshi*—Japanese for a small quantity—is pronounced (rendered in correct Japanese the name would be *Ko Tora,* or "Little Tiger"). In October 1965, the Air Force borrowed twelve combat-ready F-5As from MAP supplies and turned them over to the 4503rd Tactical Fighter Squadron (TFS), which had been formed to conduct the operational trials. The pilots underwent training at Williams Air Force Base near Phoenix, Arizona, while Northrop modified the aircraft for duty in Southeast Asia. The modifications included in-flight refueling probes on the port side of the nose, ninety pounds of belly armor plate, and jettisonable pylons underneath the wings. The aircraft were camouflaged in tan and two-tone green with light gray undersides.

At the Fighter Branch, we closely monitored the progress of the F-5 deployment. Col. Robert F. "Bob" Titus, a former Edwards test pilot, led the Freedom Fighter into combat. Titus, thirty-nine, was born in New Jersey. During World War II he was a paratrooper in the 82nd Airborne Division and attained a rank of staff sergeant. Titus earned his pilot's wings in 1949 and flew the P-51 Mustang in Korea. After graduating from the test pilot school in 1954, he spent six years in fighter test, flying all the Century Series jets: F-100, F-101, F-102, F-104, F-105, and F-106 (the F-103 was a design that was never built).

Colonel Titus and his flight left Williams on October 20, 1965, for Southeast Asia, and after multiple hookups on KC-135 refueling tankers arrived at Bien Hoa Air Base, Vietnam, on October 23. They flew their first combat mission the same afternoon. In six months, the F-5As flew over twenty-five hundred hours of close support, interception, and reconnaissance missions. Later, six more F-5As arrived, bringing the strength to eighteen. Of the original twelve aircraft, five had been lost by November 1966. Four were shot down by ground fire and a fifth crashed on takeoff because of engine failure.

The F-5 missions were exclusively over South Vietnam—they never crossed the North Vietnamese border. The aircraft did not encounter enemy MiGs, so they were denied the opportunity to demonstrate their air-to-air capabilities. Colonel Titus would later destroy three MiG-21 Fishbeds flying the F-4 Phantom.

Soon after President Lyndon B. Johnson took office following the assassination of President John Kennedy on November 22, 1963, he was briefed on a top-secret spy project. During the presidential campaign of that year, Senator Barry Goldwater continuously criticized President Johnson and his administration for falling behind the Soviet Union in the research and development of new weapon systems. Johnson decided to counter this criticism with the public release of the highly classified A-12 program. He directed that some sort of cover announcement be prepared.

On February 29, 1964, while I was a student in the test pilot school, President Johnson announced:

> The United States has successfully developed an advanced experimental jet aircraft which has been tested in sustained flight at more than 2,000 mph and at altitudes in excess of 70,000 feet. The performance of the aircraft far exceeds that of any other aircraft in the world today. The development of this aircraft has been made possible by major advances in aircraft technology of great significance for both military and commercial applications. Several aircraft are now being flight tested at Edwards Air Force Base in California. The existence of this program is being disclosed today to permit the orderly exploitation of this advanced technology in our military and commercial programs.[1]

President Johnson was talking about a secret plane that was in reality three different models of just one basic design. The Lockheed A-12 was a CIA-funded spy plane, operating under the code name OXCART, which had made its first flight two years earlier from the top-secret airfield at Groom Lake in Area 51 northwest of Las Vegas. Three YF-12A interceptor versions were later built and were also flying in Nevada. The SR-71, a Strategic Air Command reconnaissance version, would make its first flight in December 1964 from the Lockheed plant in Palmdale, California. The public didn't know the different models; they were simply all called "the Blackbird."

The president also said, "The aircraft now at Edwards Air Force Base are undergoing extensive tests to determine their capabilities as long-range interceptors."

This statement was untrue, since there were no aircraft then at Edwards and never had been. Air Force project officials were aware that some sort of public announcement was about to be made, but they had not been told exactly when. Caught by surprise, they hastily arranged to fly two of the Air Force YF-12A's to Edwards to support the president's statement. Lockheed test pilots Lou Schalk and Bill Park flew YF-12As number one, serial number 60-6934, and number two, 60-6935, from Groom Lake to Edwards. This move diverted attention away from Area 51 and the CIA intelligence-gathering nature of the A-12 project. While the A-12 OXCART program continued its secret career out at the Groom Lake site, the YF-12A performed its disguise at Edwards under a considerable glare of publicity.

A year earlier, as a student in the test pilot school, I was scheduled to fly on a Saturday morning because inclement weather had caused the school to cancel some weekday training flights. I had just started the engine of a T-33 when the Edwards control tower operator made a call on the radio.

"All aircraft hold your position," he commanded.

A few moments later I heard another transmission, "Don't look at the center taxiway."

This was a very unusual radio transmission to be made over the air. It was a strange request, an order unlike anything I had heard in my six years in the Air Force. Being a student, I followed orders. After I had flown my test mission and returned to land at Edwards, however, I happened to take a glance down Contractor's Row, the taxiway that led to all of the contractor flight operations. Parked on the ramp was the strangest looking plane I had ever seen—it was the Lockheed YF-12.

Col. Robert L. "Silver Fox" Stephens was the Air Force officer chosen to lead this top-secret program. He was nicknamed the Silver Fox because he had prematurely gray hair. Stephens, thirty-nine, was from Gilmer, Texas, and graduated from the Agricultural and Mechanical College of Texas (now called Texas A&M) in 1943 with a bachelor of science degree in aeronautical engineering. That same year he joined the Army Air Forces and flew combat, earning the Distinguished Flying Cross. After the war Stephens obtained a master's degree in aeronautical engineering from Princeton. A year after graduating from Princeton he attended the test pilot school at Wright-Patterson. In the following years Stephens flew all the Air Force jet fighters of the 1940s and 1950s era, from the F-80 to the F-106. He even flew the Bell Aircraft Company X-5 variable-sweep aircraft, the predecessor to the General Dynamics F-111.

Earlier in the 1950s his boss had asked him a question: "Want to be the Air Force test director for the Lockheed F-104 Starfighter program?" Stephens was told that the first F-104 would fly in 1954 and it might be dangerous. There was only one way to find out—fly it.

The Silver Fox accepted the challenge, one that would test his courage and his flying ability to the utmost.

Lockheed test pilot Tony LeVier made the first flight of the XF-104 in February 1954. The initial flights were routine and without incident. Subsequently, it was time for Stephens to conduct a special test on the Starfighter. To add range the F-104 was equipped with wingtip fuel tanks. If the Starfighter were to ever fly in combat it would need to be able to jettison the two tanks before engaging in a dogfight. Stephens was scheduled to make the first drop in the bombing range southeast of the Edwards main base.

"Give us a call, Colonel, after the tanks are gone," instructed the ground controller.

Stephens reached for the jettison button as he held the needle-nose jet on a straight-and-level course.

When he released the tanks he called the ground, "Tanks gone, I have just . . ."

Thump, thump!

The controls jerked in his hand, and the F-104 weaved crazily through the sky. Stephens checked his instruments quickly as he eased back on the power. He didn't know what had happened. Had the tail assembly ripped off? Was a flap gone? Had the engine thrown a turbine bucket?

"Control," he yelled out on the radio, "clear the runway and get the crash trucks. I'm coming in to land!"

"What trouble are you experiencing?" they asked.

Stephens didn't have time to answer. Very carefully he eased the engine power back and made a wide, sweeping turn, flying a flameout approach toward Runway 22 at Edwards. Still puzzled as to what had gone wrong, he delayed extending the landing gear until he was sure he could make the runway with the power now in idle. The disabled jet was still bucking when his tires squealed on the runway. Stephens managed to get the Starfighter stopped without mishap while the crash trucks surrounded him. On the ground it didn't take long for the startled test pilot to discover that the new jet needed some further modifications. When he had jettisoned the fuel tanks they had slammed into the fuselage of the F-104 instead of falling free. Only a miracle had prevented the jet from falling into a fatal dive.

After the wingtip fuel tank problem was solved, Stephens soon discovered the F-104's tendency to pitch up without warning. This fault, coupled with some other discouraging factors such as flameouts and compressor stalls in the engine and the plane's short range, sent the accident rate off the chart. Many experts in the Air Force wanted to write off the Starfighter as a failure, but Stephens wasn't one of them. He took the problems one by one and found solutions for them with the help of Lockheed engineers and other aviation consultants.

After a while Stephens' associates and superiors weren't seeing him anymore on the flight line or in his office. When other officers and pilots brought up his name, no one knew what he was doing. During his occasional public appearances he would only smile and shrug his shoulders when asked what his duties were at the time.

"Just fooling around with some planes on the flight line," he muttered, walking away.

But Lockheed's Kelly Johnson and the CIA knew differently. They had designed and were building an airplane so secret that only a handful of men in the nation knew about it. The CIA, which badly needed a successor to the U-2 after the Gary Powers disaster over Russia, provided the money for the new spy plane from the part of its budget that was not accountable to lawmakers, so there was no indication that such a project was under way. Sworn to silence, Stephens was taken into the confidence of the CIA and the few Air Force officers who knew about the project. He now lived, slept, and spent full time with the Blackbird.

The YF-12A was a monster of an aircraft, over twice the size of the Convair F-106, which was the Air Force's newest and most modern interceptor of the time. The Blackbird was 101 feet 8 inches long with a wingspan of 55 feet 7 inches and a height of 18 feet 6 inches. Maximum gross weight was 124,000 pounds, and the plane was powered by two Pratt & Whitney J58 engines, each of which produced 31,500 pounds of thrust.

The original A-12 fuselage was extended three feet for a second cockpit to house a weapon systems operator (WSO) and internal weapons bays that would carry three Hughes Aircraft Company GAR-9 air-to-air missiles (later designated the AIM-47). The YF-12A differed from the A-12 in that it had a folding ventral fin under the rear fuselage to add directional stability due to a more rounded nose to enclose the Hughes ASG-18 radar. The cockpit in the YF-12A was slightly raised when compared to the A-12, so the pilot could see over the rounder nose when the aircraft was taking off and landing. It also had a ball-shaped infrared sensor on each side of the chine below the cockpits and carried streamlined camera pods under the engine nacelles to photograph missile launches.

With the completion of the missile launches Air Force planners calculated that Air Defense Command would require ninety-three F-12B advanced interceptors to provide a defensive screen that could protect the entire United States from Soviet bombers during the Cold War.

After President Johnson made his dramatic disclosure in 1964 the world was buzzing about the plane. But strict secrecy continued to be clamped on the actual facts and performance of the aircraft. Experts all over the world argued about the aircraft and especially whether President Johnson's statement was propaganda or fact. Was it actually a Mach 3 plane? Could it cruise at 70,000 feet? If true, the YF-12A Blackbird couldn't be intercepted by any Russian fighter and it was doubtful that any enemy ground-based missile system could down the aircraft.

While I was at Test Operations in April 1965, Stephens got the word that it was time to back up the president's words with action.

"Many people, especially the Russians, don't believe we have an aircraft capable of reaching the speed or altitude which was announced," he was told by a senior Air Force official. "It's up to you to prove that the YF-12A can do what we said it can do."

"Officially?" Stephens asked, thinking the flight should be conducted in secret.

"Roger," said the official, "we want you to go after the world speed and altitude records which the Russians established in 1962 in their E-166 jet."

The day chosen for the flight was May 1, 1965. Stephens selected Lt. Col. Daniel Andre to go with him as his WSO. The two men sat through a pre-mission briefing and realized that this was no ordinary test hop. They checked the weather and went over the special characteristics of the flight. There was no question that Stephens wanted everything perfect for this record try, that he didn't want any mishap to mar the attempt to beat the Soviets.

Stephens and Andre suited up in full-pressure suits, the same suit used by NASA Gemini astronauts. Minutes later the Silver Fox lifted the all-black plane number three YF-12A, serial number 60-6936, off the runway at Edwards and headed for the sky and aviation history. Below him on the ground, watching his every maneuver, were military and civilian test pilots, engineers, technicians, and other experts who belonged to Stephens' test unit. It was rumored that President Johnson kept a line open to learn the results of the record-breaking attempt.

The speed record try was made over a seventeen-kilometer straightaway course in opposite direction runs. Stephens, as he turned his aircraft toward the course, kept repeating the speed he had to beat, the record held by the Soviets, "Sixteen hundred and sixty-five mph . . . 1,665.89 mph."

Then, like a flash of light, he was on the course.

"Starting run," Stephens called to the ground controllers.

Moments later they called out, "Eighteen hundred mph."

The radio call came through Stephens' headset weakly, as though the controller was at the other end of the United States. But Stephens heard it and understood that he needed more power from the J58 engines and afterburner if he was to best the Soviet record by a significant amount.

"Nineteen hundred mph," was the next call.

Stephens pushed the throttles further open, watching his instruments carefully. The sleek YF-12A was still increasing in speed.

"Twenty hundred and seventy mph!" shouted an elated controller.

A grin creased the sweating face of the Silver Fox as he heard the radio call. That was one record for the United States, he thought.

His assault on the cruising altitude record was just as successful. The Russians had held an absolute height of 74,376 feet in the E-166 jet. Stephens and Andre calmly took the YF-12A to an absolute altitude of 80,257 feet and cruised back and forth for the FAI (Federation Aeronautique Internationale) timers on the ground below.

On the same day, Maj. Walter F. Daniel and Maj. Noel T. Warner set a 500-kilometer closed-circuit speed record in the YF-12A of 1,642.042 mph. Daniel flew again that day with Capt. James P. Cooney and set a 1,000-kilometer closed-circuit speed record of 1,688.891 mph. These four records were passed on to the press and spread around the world. It was obvious that the YF-12A had no equal.

Asked later by the press whether he had flown all out on the flight, Col. Robert Stephens shrugged, "If the Russians or any other nation breaks our records, I think we will go back up and see what we can do about it."

Maj. Frank Liethen, the pilot who gave me my first flight in the F-104 Starfighter, danced to his own drumbeat. From Wisconsin, he danced the polka. Polka music was not played in the mid-1960s at the Edwards Officers Club, where rock and roll was king of the dance.

Liethen, five years my senior, was six feet three inches tall and weighed 220 pounds. Generally quiet but quick witted, he nearly always had an unlit cigar in his mouth. He was very intelligent and well educated—you felt a presence when you were in Liethen's company. He was a man's man, the last person you would picture doing the polka on the dance floor.

Regardless of the song being played on the jukebox or by the band, Liethen, nicknamed the Dancing Bear, would grab my wife Jan and dance the polka. Other dancers gave them wide room to perform. Pilots at Edwards in the 1960s set the pace on the ground and in the air. Frank Liethen was a leader, definitely not a follower.

After Yeager's accident and the loss of one NF-104, only two aircraft remained. As a result of the accident investigation, several modifications were made to the aircraft and operating procedures. The aircraft underwent further testing at Test Operations during 1964 while I attended the school. No one in our class got to fly the NF-104—it was a great disappointment. After mission procedures were altered and the NF-104 pilot's handbook upgraded, the two remaining aircraft were delivered to the school in early 1965. By then I had been assigned to the Fighter Branch of Flight Test Operations and continued to fly the basic Starfighter.

At the school the NF-104s were placed under the operational control of Maj. Robert H. "Bob" McIntosh, Aerospace Division operational chief, and his assistant, who was Frank Liethen. Bob McIntosh had been a F-86 Sabrejet pilot in the Korean War (one MiG-15 destroyed, three damaged) and a demonstration pilot with the Air Force Thunderbirds for two years as a solo pilot and left wing.

McIntosh and Liethen wrote the school syllabus and were prepared to teach students about pressure-suit operation and the ground and flight handling of the NF-104. Both pilots started to fly the aircraft to acquaint themselves with its flight characteristics.

On June 18, 1965, Liethen was scheduled to fly a zoom training mission in NF-104, serial number 56-0756, using a call sign of Zoom 1. Navy Lt. Jack Finley, one of my classmates from test pilot school, was assigned duty as chase pilot in another F-104 using the call sign Chase 1. Liethen flew about one hundred nautical miles east of Edwards, reversed course, and started to accelerate in speed by using his afterburner. He then lit the rocket engine at an altitude of 35,500 feet and on reaching Mach 2.03, started a 3.5 G pull-up to a climb angle of 45 degrees. Liethen flew a precise profile, experimenting with the reaction controls as he peaked at an altitude of 98,850 feet. The flight proceeded as planned and he started to descend back toward a landing at Edwards. After he relit the J79 engine he told Finley he could break off.

Passing through 20,000 feet a few miles northwest of Edwards, Liethen felt a jolt and heard a thump in the aircraft. He later described it as being similar to an incident years earlier when a crow had struck his canopy while he was flying a F-86.

Liethen transmitted on his UHF radio, "Jack [Finley], are you still with me?"

"Negative," responded Finley. "Do you need me?"

Liethen answered, "There was a pretty violent knock in the airplane," and then said, "Eddie tower [Edwards Air Force Base control tower], Zoom 1."

"Zoom 1, Edwards tower," answered the tower operator.

Since all cockpit gauges and indicators were normal, Liethen set up a precautionary landing pattern to the Edwards 15,000-foot main runway.

"Emergency," called out Liethen, "I am going to land 04 [Runway 04]. I am on the high left base leg."

"Zoom 1," directed the tower operator, "winds calm, altimeter two-niner-niner-seven, report gear down, cleared to land."

Within twenty-five seconds Jack Finley joined up in a loose trail position on the left side of Liethen's stricken NF-104, but he could not see any damage.

"I got you," called Finley. "I am with you now."

"It might have been just a compressor stall," Liethen explained to Finley, and then to the tower he said, "Short final."

"Zoom 1, cleared to land," directed the tower.

In just one minute and thirty seconds after the mysterious explosion, Liethen safely landed the Starfighter and let it roll to the end of the runway. Taxiing back to the ramp on the east taxiway, Liethen waved to the pilot of another F-104 and he waved back. Liethen cancelled his emergency. Three fire trucks followed him toward the NF-104 recovery area.

As he taxied by the control tower, the operator informed Liethen that his emergency was not over:

"Zoom 1, are you aware that you appear to be siphoning fuel or smoke from your right wing?"

Liethen twisted around in his pressure suit, observed smoke streaming from the right wing, and immediately proceeded as fast as he could taxi to the NF-104 refueling pad for emergency dumping of hydrogen peroxide. Five minutes had passed since Liethen had first heard the explosion.

"I will get this thing on the pad as quick as I can," replied Liethen.

"It appears to be burning," the tower operator informed him. "It appears that your right wing is burning."

"Tell the fire trucks to catch me," said Liethen. "I have to put this thing on the pad."

In the recovery area, Liethen evacuated the NF-104 without injury and the fire department extinguished the fire. Extensive damage occurred to the right wing and tail section. The incident was considered a major accident, and I was selected to be the official accident investigator.

It was not difficult to determine what had caused the mishap. Leaking hydrogen peroxide from the reaction control system, the same system used in the X-15, caught fire and exploded. The explosion caused extensive fire damage in the right wing-root area and blew the specially modified drogue chute and several pieces of aircraft skin off the NF-104. A slip joint connecting the hydrogen peroxide line from the fuselage to the wing had failed because two rubber O-rings had been damaged on installation two years earlier. Both O-rings were removed and analyzed by the base laboratory. They were scarred and lacked lubrication to make a good seal.

We found the drogue chute about five to seven miles northwest of Edwards, exactly where Liethen had described his position when the explosion occurred. The accident board, made up of myself and six other Air Force officers, thought the investigation would be more complete if we could also recover several small pieces of aircraft skin that departed the NF-104 during the explosion. After writing the description of the aircraft, the accident sequence, and results of the preliminary investigation, my next job was to find the missing parts.

Capt. David H. "Dave" Thomas, thirty-two, a classmate of mine in the test pilot school, was transferred to the Helicopter Branch of Test Operations. Thomas was an easygoing type and had just married before attending the school. He was prematurely bald and extremely interested in helicopter testing. Like me, his job was to support the major programs at Edwards such as the X-15 and XB-70.

On a hot summer afternoon in late June, Dave Thomas piloted a twin-rotor Piasecki H-21 Flying Banana helicopter over the Mojave Desert with me as his copilot. Thomas and I looked for several small pieces of aircraft skin, the largest about two by three feet. We figured the lightweight skin would have landed several miles to the east of the drogue chute due to the prevailing wind. Thomas set up a pattern and crisscrossed the desert at about 80 knots and around 200 feet above the desert floor. The helicopter was not air conditioned so the heat soon became unbearable. Also, I developed an instantaneous headache from the flicker of the sun shining through the whirling rotor blades. Back and forth we flew looking for the proverbial needle in a haystack. Our best hope was that the sun would reflect off the aluminum skin and attract our attention, but the desert was filled with trash such as discarded bottles, cans, and hiking refuse. Thomas and I gave it the old college try but in the end came up empty-handed. He did let me fly the helicopter for quite some time—just in horizontal flight, not in the vertical mode. After my first flight in a helicopter, I was glad I had decided to test fighters.

Our official accident report stated that the damage to the NF-104 was minimal and could be repaired in a few weeks for about $15,000. Fortunately, no one was injured or killed in the accident. Of great concern to our board was the lack of procedures in some instances, the disregard of procedures in other instances, and the general lack of safety awareness by people involved in the accident. The board thought there were a lot of "ifs" in the accident:

- If the two O-ring seals had been installed correctly there would not have been an accident.
- If regulations had been followed that called for O-ring seals to be removed and inspected, the error might have been caught.
- If Liethen had kept Finley as a safety chase, even after his zoom, he might have had better information while airborne.
- If Finley had joined on Liethen's right wing he would have seen the external damage.

- If Liethen had not canceled his emergency after landing, he would have been given more assistance.
- If the control tower operator had radio or phone connection with the NF-104 recovery area, the ground crew would have been prepared to put out the fire faster.

Frank Liethen had been lucky the explosion occurred near the base and not one hundred miles east of Edwards during his turn-in. He was fortunate that most of the hydrogen peroxide had been consumed during his zoom maneuver. He used good judgment in landing the aircraft as soon as possible, which allowed him to arrive in the recovery area in just seven minutes. But the accident was another close call for the Air Force Flight Test Center, a harbinger of bad events to come both for the test center and for Frank Liethen.

Bad news travels fast. The commandant of Test Operations received a military Teletype message (TWX, or Teletype wireless exchange) reporting that one of our pilots had been killed in Europe. All the fighter, bomber, cargo, helicopter and VTOL (vertical takeoff and landing) test pilots had offices in the same building. So within a short time we had all gotten to know one another.

Maj. Philip E. "Phil" Neale Jr., thirty-five, had been killed in the crash of a French Dassault "Balzac" VTOL research aircraft he was testing. The mishap occurred at the Centre d'Essais en Vol, the French flight test center at Bertigney, fifty miles south of Paris. At Edwards we tested all the aircraft manufactured in the United States that were purchased by the Air Force. Sometimes pilots traveled abroad to evaluate aircraft built in foreign countries.

Phil Neale and I had flown a T-33 together. Although I did not know him very well, I did know he must have been an accomplished pilot to be assigned to fly VTOL aircraft. Neale had not always been a VTOL pilot. He came to the test pilot school in 1959 as a student after spending three years in Bitburg, Germany, as a fighter pilot and maintenance officer in the 525th Fighter Interceptor Squadron. Then he spent two years flying the F-86 Sabrejet with the 93rd Fighter Interceptor Squadron at Kirtland Air Force Base, New Mexico.

After graduating from the test pilot school, Neale remained as an instructor and taught aircraft performance testing. He was a fun-loving type who worked hard at his job and then partied hard in the evening. Phil Neale was married and had five children. His classmates at the school selected him for the Propwash Award given to the student who best kept their morale up throughout what was then an eight-month course.

While instructing at the school, Neale had a close call. He was flying the backseat of the T-33 with a student in the front seat when the canopy blew off. Despite windblast and subzero temperatures, Neale flew back to Edwards and landed safely. Lt. Col. Richard Lathrop, the commander of the school at the time, was very impressed with Neale's flying skills and also his teaching ability.

The school received a special request from the Army, and Colonel Lathrop had so much confidence in Neale that he turned the assignment over to him. At the time, the Army did not have a test pilot school or flight test facility. They requested that the Air Force test pilot school conduct a special class for seven of their aviators using a Bell UH-1 Huey. Neale had never flown a helicopter before, so he checked out in an Air Force Huey at Hill Air Force Base, Utah. With little chopper experience, he set up the academic curriculum and personally trained all the Army aviators.

With his new learned experience in helicopters, in January of 1965 Neale transferred to the VTOL Branch of Test Operations. Barely three weeks before his fatal flight in the Dassault Balzac, he became the first Air Force pilot to fly the Ryan XV-5A Vertifan research aircraft, which was undergoing tests at Edwards. The XV-5A, like the Balzac, was a dangerous machine and had already killed one pilot.

On April 27, 1965, the first XV-5A, serial number 62-4505, crashed at Edwards, killing Ryan Aeronautical Company test pilot Louis "Lou" Everett. Everett, forty-one, was born in Brooklyn, New York. During World War II he was too young for the Navy cadet program, so he enlisted in the Army. Within a few months he transferred to the Army Air Forces and began training as a fighter pilot assigned to fly P-51 Mustangs. Stationed in Florida, he was awaiting assignment to go overseas when the war ended. Everett was recalled into the Air Force to serve in Korea, where he flew AT-6 Texans on forward air control missions. He returned to the States to resume his education at Mississippi State, graduated in 1954 with a degree in aeronautical engineering, and joined Chance Vought in California as an engineer. However, Everett still yearned to fly. Later he was hired by Ryan Aeronautical Company as their second test pilot for the X-13 Vertijet, joining Ryan's chief test pilot, Pete Girard. The X-13 was the world's first pure jet VTOL aircraft, and Girard and Everett were the only pilots to fly it.

Lou Everett took off in the XV-5A in the vertical mode and flew southeast over Rogers Dry Lake where he made a 270-degree turn and conversion to conventional flight. He made a low pass on a southwesterly heading near the crowd at an airspeed of 310 knots and an altitude of 50 feet, ending with a sharp pull-up in a left climbing turn. The aircraft then decelerated prior to jet-to-fan conversion on an easterly heading. Item four on the checklist, a check of the flight-control electrical switch, was interrupted with words that sounded like "I've got to get out." At the time the XV-5A was at an altitude of 800 feet, two miles from a group of three hundred reporters and military and civilian guests who had been invited to watch the demonstration. Everett ejected at low altitude as the aircraft dived toward the lake bed at an estimated 30- to 40-degree angle. The XV-5A was in the preconversion mode, flying at approximately 140 knots. Full conversion would have been completed two minutes later at about 90 knots.

The accident board speculated that Lou Everett inadvertently hit the conversion switch at too high an airspeed. The conversion switch was a simple two-position toggle switch located on the collective for pilot convenience. During conversion from horizontal jet flight to vertical fan flight the pilot activated the conversion switch, which caused a programmed configuration change that included opening the wing fan doors, opening the nose louvers, and activating fans and reaction controls. Most important, though, it adjusted the stabilizer to a full-down-minus-15-degree position. Activation of the switch at high speed caused an immediate nose down rotation of the aircraft with negative G forces, throwing the pilot's arms away from the controls as the XV-5A dove into the ground. Unfortunately, Everett's ejection seat was improperly rigged and he was instantly killed.

Lou Everett and Phil Neale were the first of many test pilot friends I would lose during my years at Edwards. I had only known Phil Neale for a few months, but like the other test pilots at Test Operations, I was shocked and saddened to hear he had been killed in a similar fashion to Lou Everett.

I asked the other test pilots if they knew of any living Air Force pilot who had flown two different VTOL-type aircraft. No one could think of a name—none had survived. The VTOL aircraft were generally built as one or two of a kind and then first tested by the manufacturer's test pilots. Because so many VTOL aircraft crashed and killed their pilots, flight experience in this radically new type of flying was severely lacking.

Dassault Aircraft Company, builder of the Balzac that killed Phil Neale, was already a very successful builder of French fighter aircraft before they ventured into VTOL aircraft. In first building a fighter-type aircraft they opted for a slender delta wing in their design and had flown the tiny Mirage I in June 1955. This design grew into the Mirage III, a 15,000-pound Mach 2 single-seat fighter, by late 1956.

To gain VTOL experience, Dassault started their design with an existing airframe and available engine. The airframe was that of the Mirage III prototype, III-001, and the engines were Rolls-Royce RB108s that had previously been used by a British company called Shorts of Belfast. Shorts installed the engine on a tiny delta-winged VTOL aircraft, the Short SC1. Two aircraft were built and flown in the late 1950s. Although one crashed because of gyro problems, killing the pilot, the Rolls-Royce engine performed well.

Dassault built a half-size scale model of their proposed final design, which weighed 15,000 pounds. They designated it V001, V standing for vertical and 001 being the airframe number. The nickname "Balzac" derived from the telephone number of a well-known advertising agency, Balzac 001.

The Balzac retained the general features of the Mirage and had a 24-foot slender delta wing with 60 degrees of sweep, a large, sharply swept tail fin, and a slim pointed nose with single-seat cockpit set well forward and semicircular engine air intakes on each side. The fuselage was lengthened to forty-three feet, and the center section was made taller and much wider. Eight 2,160-pound RB108 lift engines were fitted in two parallel rows of four. The rows were separated to allow the intake ducting for the propulsion engine to pass between them, and in each row the front two engines were separated from the rear two by the main gear retraction bay. With the lift engines taking up most of the center fuselage, there was no room for the original 10,000-pound-thrust Atar engine, so a much smaller 5,000-pound-thrust Bristol Orpheus engine was fitted into the remaining space in the rear fuselage.

With test pilot René Bigand at the controls, the Balzac made its first tethered hover on October 12, 1962, followed by a free-flight hover on October 18. After completing hovering trials, it made a conventional all-horizontal flight on March 1, 1963. It then progressed rapidly to its first complete cycle—vertical takeoff, conventional horizontal flight, and vertical landing—on March 29, four months ahead of schedule. The Balzac continued flying until, on its 125th flight on January 27, 1964, it crashed during roll testing, killing Jacques Pinier. The airframe was salvaged, so V001 returned to flying status once again.

In September 1965, Phil Neale was on temporary duty (TDY) in Europe as a member of an Air Force evaluation team assigned to test the Balzac. During hover tests, the Balzac ran out of fuel and the engine quit. Either Neale misread the fuel-quantity gauge, which was marked in French, or the remaining fuel was in the wrong tank. He was at low altitude and made the decision to eject. However, due to human error, the ejection seat was in the high-speed mode, not the zero-zero altitude mode used for hovering. As a result Neale's ejection did not give him enough altitude for full parachute deployment. Phil Neale was instantly killed when he hit the ground.

The single Balzac VTOL aircraft that was built gained the rare and unhappy distinction of killing two pilots in separate accidents.

I flew routine chase flights for both other Air Force pilots and civilian contractors in between my own test missions in the Fighter Branch at Edwards. Chase pilots flew a loose formation position about fifty to one hundred feet abeam the test aircraft, and looked for fuel or oil leaks, smoke trailing behind any part of the plane, missing side panels, or any other unusual conditions and called them to the attention of the test pilot. We would also be an extra set of eyes helping to prevent a midair collision.

On September 25, 1965, a day after my thirtieth birthday, I was flying a T-38 Talon on a safety chase mission when I heard a disturbing call on the radio.

"Mayday—Mayday—Mayday! This is N1038 Victor, the Super Guppy, on a flight test over the Mojave Desert. We have had a major structural failure of the upper nose section during a maximum speed dive and are preparing to bail out!" said the excited voice.

At the time the Super Guppy, registered by the FAA (Federal Aviation Administration) as N1038V but called the 377SG, was the world's largest airplane. It was an antique Boeing C-97 Stratocruiser that had been modified by Aero Spacelines of Van Nuys, California, to carry the Douglas-built S-IVB third stage of the Saturn V launch vehicle and the lunar excursion module (LEM) built by North American Aviation, both of which were too large to be transported by existing aircraft. The standard cargo section of the Super Guppy was increased to about 50,000 cubic feet, approximately five times the size of a Boeing 707 transport, by enlarging the fuselage from 8 feet 10 inches to a cavernous 25-foot diameter. Four Pratt & Whitney T-34 turboprop engines, each capable of producing 7,000 total equivalent horsepower, powered the Super Guppy.

"N1038 Victor," responded the Edwards control tower operator. "May we help you?"

"Stand by," said the Super Guppy's civilian test pilot. "We have a large hole in the nose and the aircraft is disintegrating and buffeting severely. N1038 Victor will advise intentions."

The Super Guppy pilot had completed all the tests required for airworthiness certification except the most hazardous maneuver—the high-speed dive. This portion of the tests was saved for the last phase and called VD (velocity dive). The test crew reported a tremendous bang and a jolting, violent shudder as the huge cargo ship hit 275 miles an hour at 10,000 feet. They thought they had had a midair collision. The Guppy wallowed like a giant Moby Dick plowing through mountainous waves, but she managed to fly.

The high-speed dive had to be flown at the maximum gross takeoff weight of the aircraft. The crew had prided themselves in being resourceful and had arranged to borrow thirty thousand pounds of borate in one-hundred-pound sacks from a chemical dealer in Mojave. As the tearing, shredding metal from the nose blew aft inside the mammoth interior, the flying slivers ripped holes in the paper sacks of borate powder. The whole interior, including the cockpit, was filled with a swirling cloud of powdered borate by the slipstream being rammed into her.

Then one of the crew members yelled out, "The whole damn nose section has caved in. We've got a hell of a hole over the cockpit!"

The plane had a gaping twenty-three-foot hole, and pieces were still collapsing and tearing off. Broken stringers and pieces of frame were being ripped loose, shooting through the air like arrows, impaling themselves like steel through tinfoil in the frames that supported the fuselage at the rear of the cargo compartment. It was like flying a giant scoop.

But just when it seemed as if they had absolutely nothing going for them, Lady Luck smiled. The inspection hatches and access doors blew out in the tail section, and this relieved the immense internal air pressure that had been threatening to blow the ship apart. The haze of borate also was whisked away by the rushing air.

"Thirty-eight Victor," called Edwards tower, "we have a DC-9 in flight test on takeoff roll. Can he help you?"

"Roger, Edwards," replied the Guppy pilot. "We'll take the DC-9."

I called Edwards tower from my plane and offered my assistance, but they turned me down since the DC-9 was closer to the Guppy. Soon the DC-9 pilot joined up on the Guppy's right wing. The Guppy pilot asked for his status.

"Difficult to tell," said the DC-9 pilot. "We can't get too close yet, but pieces are flapping and still coming off the thing."

When the DC-9 was finally able to make a visual sweep of the Guppy's tail, he gave an encouraging report.

"The tail appears to be intact with no apparent damage," he said.

It was then the Guppy pilot made a decision to try to save the damaged aircraft. He maintained 175 mph airspeed and began a shallow descent toward Edwards for a landing attempt on the adjacent Rogers Dry Lake bed.

"Edwards tower, Thirty-eight Victor, ten miles north at 4,000 feet. Request landing on the dry lake."

"Roger, Thirty-eight Victor," responded the tower. "Land to the south. Crash equipment is positioned and standing by."

The Guppy's approach to the dry lake was long and flat, and they didn't use wing flaps. The touchdown was smooth and they let the Guppy roll out to a dead stop on the long lake bed, and then taxied to the Edwards ramp and shut down.

Author George J. Marrett, seven (left), and childhood friend Bob Preston, eight, (in 1943) are ready to fly during World War II. (*Author collection*)

Second Lt. George J. Marrett climbing into Lockheed T-33 T-Bird during USAF flight school at Webb AFB, Texas, 1958.

Front row (l-r): 1st Lt. Lenard W. Kresheck (instructor), 2nd Lt. George J. Marrett (student). Back row (l-r): 2nd Lt. David G. Mosby (student), 2nd Lt. Jack D. Brooks (student) at Webb AFB, Texas, 1958.

Capt. George J. Marrett standing in front of McDonnell Aircraft F-101B Voodoo at Hamilton AFB, California, 1963.

Maj. William A. Anders, NASA astronaut, 1963.

Pancho Barnes and Col. Charles E. "Chuck" Yeager, 1964.

Col. Charles E. "Chuck" Yeager in front of the tail of Lockheed NF-104 Starfighter, serial number 60-760, at Edwards, 1963.

USAF Test Pilot School Class 64A at Edwards AFB, 1964. Front row (l-r): Capt. Dick Voehl, Capt. Dave Livingston, Capt. Jerry Tonini, Lt. Jack Finley (USN), Capt. Jim Hurt, Capt. Dave Thomas, Lt. Dick Truly (USN), Capt. Bob Whelan. Back row (l-r): Fred Haise (NASA), Capt. Harry Wilkerson (Canada), Capt. Ron Franzen, Don Mallick (NASA), Capt. Ernie Hasselbrink, Capt. Jap Hofstra (Netherlands), Capt. Dick Strickland, Capt. George Marrett.

Capt. George J. Marrett kneeling in front of Lockheed F-104A Starfighter at Edwards AFB, 1965.

Capt. Frank E. Liethen Jr. sitting in the cockpit of Lockheed F-104A Starfighter at Edwards AFB, 1964.

Capt. Charles "Chuck" Rosburg sitting in the cockpit of Lockheed F-104A Starfighter, 1964.

Test pilots of Flight Test Operations. Front row (l-r): Capt. Al Hale, Maj. Merv Evenson, Maj. Fred Cuthill, Capt. Ron Franzen, Maj. Gus Julian, Maj. Mike Adams, Capt. George Marrett, Capt. Dave Livingston, Maj. Bill Reisch, Capt. Jerry Gentry, Maj. John Ludwig. Back row (l-r): Maj. Charlie Bock, Capt. Bill Frazier, Maj. Russ Bunn, Maj. Ivan Skinner, Maj. Fred C. Dobbs, Maj. Ted Sturmthal, Capt. Floyd Stroup, Capt. Jerry Bowline, Maj. Joe Schiele, Capt. Jerry Tonini, Capt. Tom Smith, Capt. Bob Whelan, Lt. John Miller (RAF), 1967.

Capt. James H. "Jim" Hurt III on ladder of Republic F-105 Thunderchief at Kirtland AFB, New Mexico, 1965.

Rockwell test pilot Tommie "Doug" Benefield standing in front of B-1 Lancer, 1981.

Aero Spacelines Super Guppy damaged on high-speed dive at Edwards AFB on September 25, 1965.

Col. Robert L. "Silver Fox" Stephens standing in front of Lockheed YF-12 Blackbird after setting a world speed record on May 1, 1965, at Edwards AFB.

Capt. James E. "Jim" Taylor standing in front of Lockheed F-104A Starfighter at Edwards AFB, 1964.

Maj. Philip E. "Phil" Neale standing in front of Piasecki H-21 Flying Banana at Edwards AFB, 1963.

Capt. George J. Marrett flying chase on North American XB-70A Valkyrie, 1965.

Maj. Carl C. Cross, 1965. Killed in XB-70/ F104 crash.

NASA test pilot Joseph A. "Joe" Walker pre-flighting Lockheed F-104B Starfighter at Edwards AFB, 1965. Killed in XB-70/ F-104 crash.

North American Aviation XB-70A Valkyrie flying over Edwards AFB, 1965. This was the number two aircraft lost in the midair crash.

Capt. George J. Marrett standing in front of Northrop T-38A Talon at Edwards AFB, 1964.

General Dynamics F-111A (serial number 63-9774) wet-runway tests at Edwards AFB. Back row: all unidentified. Front row (l-r): Capt. George J. Marrett, Maj. Robert "Bob" Parsons, two unidentified, 1966.

General Dynamics F-111A Aardvark, serial number 63-9774, after it crashed on January 19, 1967, at Edwards AFB, killing Maj. Herbert L. "Herb" Brightwell.

Capt. George J. Marrett standing in front of McDonnell F-4C Phantom, serial number 63-7409, at Edwards AFB, 1967.

McDonnell Aircraft F-4C Phantom, serial number 63-7409, test aircraft at Edwards AFB, 1967.

Capt. Lawrence C. "Larry" Curtis Jr. on ladder next to Lockheed F-104A Starfighter at Edwards AFB, 1967.

Capt. Joseph F. "Joe" Stroface sitting in the cockpit of Lockheed F-104A Starfighter at Edwards AFB, 1965.

Maj. David H. "Dave" Tittle, 1965. Killed in XV-5A crash.

North American Aviation test pilot Van Shepard, 1965. Killed in Mini Guppy Turbine crash.

Maj. Robert A. "Bob" Rushworth standing in front of North American Aviation X-15 at Edwards AFB, 1965.

NASA test pilot Bruce A. Peterson after piloting the first flight of the Northrop HL-10 lifting body at Edwards AFB, 1966.

Maj. William J. "Pete" Knight kneeling in front of North American Aviation X-15 at Edwards AFB, 1967.

Maj. Robert H. "Bob" Lawrence standing in front of Lockheed F-104A Starfighter at Edwards, 1967.

Maj. Michael J. "Mike" Adams standing in front of North American Aviation X-15, 1967. Handwriting reads, "To 'George Marrett' An outstanding pilot and fine friend. Mike Adams Major, USAF."

Maj. Robert M. "Bob" White in cockpit of North American X-15 after setting world altitude record, 1961.

After my mission I taxied by the Super Guppy and took a 35mm color slide with my Argus 3C camera that I carried in my helmet bag on all chase flights. The Guppy was repaired and later carried the lunar excursion module *Eagle* and command module *Columbia* flown by *Apollo 11* astronauts Neil Armstrong, "Buzz" Aldrin, and Mike Collins in their moon landing of July 1969. For seventeen minutes the Super Guppy's future looked uncertain, but, due to superb piloting and Rogers Dry Lake, it survived to make aviation history.

On October 6, 1965, I flew a routine flight in an F-104, followed by another routine flight in a T-38 the next day. But what happened at Edwards the night of October 6–7 was anything but routine. About an hour after midnight SSgt. Charles A. "Chuck" Sorrels, a control tower operator assigned to the 1925th Communications Squadron, spotted some unusual lights in the sky. Sergeant Sorrels contacted SSgt. Forrest S. Honer, a base operations dispatcher, who verified the sightings of the strange lights. Sergeant Sorrels estimated there were from seven to twelve objects involved. The objects had the general shape and appearance of stars, with the largest about twice the size of an ordinary star. The others were the same size as stars, which made them about the size of a pinhead when held at arm's length. These mysterious lights appeared to flash red, green, and white at one time or another. Sergeant Sorrels could not discern a formation with the exception of three objects that appeared to be in what might roughly be described as an inverted "V." All the objects were first observed fifty miles west of Edwards and traveled to a position about forty miles south. They never entered Restricted Area R-2515, the area in which all of our flight testing took place. The incident would later be termed by those who believe in UFOs "the Edwards Air Force Base UFO Encounter."

Capt. John F. Balent, tactical evaluation officer from the Foreign Technology Division at Wright-Patterson Air Force Base in Dayton, Ohio, was asleep in the base housing area. I was also asleep at home. Captain Balent was contacted by phone and arrived about 3 AM to witness this strange phenomenon and make an official report. Captain Balent reported the activity level had subsided by the time he arrived, but Sergeant Sorrels still claimed to see flashing lights on two of the four objects.

No Flight Test Center aircraft were on alert at Edwards to take off and identify the source of the lights. However, six nuclear-armed F-106 Delta Darts from the 329th Fighter Interceptor Squadron at George Air Force Base near Victorville, California, were on alert sitting on the Edwards ramp.

Over the next few hours Edwards officials contacted the Los Angeles Air Defense Sector (LAADS) of Air Defense Command (ADC), the controlling authority for the interceptors, and requested aircraft be scrambled to determine the identify of the unidentified flying objects. Officers at LAADS contacted their headquarters at the 28th Air Division at Hamilton Air Force Base, California. Just three years earlier, during the Cuban Missile Crisis, the headquarters at Hamilton was also faced with an extremely difficult decision. At that time they wanted to disperse their interceptors to civilian airfields and keep them out of the range of Soviet ICBMs stationed in Cuba. I had been a part of that fiasco, since our headquarters never expected nuclear-armed interceptors to be flown unless the United States was at war.

During the night of October 6–7, 28th Air Division officers determined that the UFOs were not hostile and therefore not a threat to the United States. As such, regulations prevented scrambling nuclear-armed interceptors. It was decided to call in a pilot from the 329th FIS and launch him into the night sky over Edwards in an unarmed F-106.

Capt. Darryl Clark took off about 5 AM and Sergeant Sorrels vectored him to a position to chase the moving lights. Both Edwards tracking and weather radar picked up the unidentified objects and determined they were at a relatively slow horizontal speed but able to climb at a tremendous rate. The Edwards height-finder radar plotted the objects climbing from 3,000 feet to 13,000 feet in just a few seconds. Captain Clark spotted one light and climbed to 40,000 feet but still could not identify the strange object. All the while voice recordings were made between the pilot, the control tower operator, and Air Defense Command officers.

Eventually, the sun came up and all the mysterious objects disappeared. On an audiotape that recoded the incident, the observers discussed possible explanations for what they had just seen. No one had ever seen lights like this in the past and could not offer any answers.

Finally, one person said, "I guess we'll never know."

When I arrived at Test Operations the next morning rumors were circulating about the unknown events of the previous evening. Most pilots were lighthearted about the story and laughed it off. No one believed the UFOs were aliens flying vehicles from outer space. Only a year and a half earlier two YF-12s had arrived at Edwards on an unannounced flight from Groom Lake northwest of Nellis Air Force Base, Nevada. It was entirely possible that other classified projects were operating at night over and around Edwards. If so, the pilots could well be assigned to Test Operations and would be under strict orders to not identify themselves or to talk about their aircraft. If there were a new class of high-performance aircraft and pilots flying them, I wished I were one of them. Eventually, talk about the UFOs died down; we had more important activities to pursue.

Chapter 9

Fighter Branch

Winter mornings dawned beautifully on the Mojave Desert. The temperature was usually in the low 30s, with visibility unlimited and calm winds. It was a great time and place to test airplanes.

Early on a January morning I checked in to our Test Operations facility, donned a flying suit, and reported to our flight dispatcher. He outlined my mission, a contractor chase for a Lockheed twin-engine SR-71 Blackbird. I would fly an F-104 Starfighter, the little brother of the Blackbird. Lockheed's brilliant aircraft designer Kelly Johnson designed both the SR-71 and the F-104. The F-104A was a top air-to-air fighter in the late 1950s and early 1960s, and the SR-71 was the newest Mach 3.2+ spy plane in the Air Force inventory.

The SR-71 was the latest and best of the Mach 3 family. It had the same wingspan and tail height as the YF-12A, but its length was about six feet longer. With additional fuel on board, the SR-71 had an unrefueled range of about three thousand miles, significantly greater than the YF-12A. Only three YF-12As were built, but they never went into production. On the other hand, thirty-two SR-71s were built, becoming operational in a strategic reconnaissance wing at Beale Air Force Base, California, in January 1966. The aircraft remained in service for twenty-four years and finally retired because of a decreasing defense budget and lack of spare parts.

My mission that day would take only thirty or forty minutes starting shortly after sunrise. Robert J. "Bob" Gilliland was the Lockheed pilot who would perform the tests on the SR-71 using the call sign of Dutch 51. Gilliland, thirty-eight, from Memphis, Tennessee, had joined Lockheed five years earlier as a production test pilot on the foreign versions of the F-104. He helped train the first German pilots to fly the Starfighter as well as pilots from Italy, Holland, Belgium, Canada, and Japan. In 1962 he returned to the United States and joined Kelly Johnson's Skunk Works as one of the first pilots to fly the YF-12. Gilliland then flew the first flight of the SR-71 in December 1964.

I first met Gilliland at one of Lockheed's annual Christmas parties in Burbank, California. Company test pilots Tony LeVier, Herman "Fish" Salmon, and Darryl Greenamyer hosted other contractor pilots in the Los Angeles area and a few of the military pilots stationed at Edwards to a suit-and-tie dinner in the company's dining room. This dinner traditionally started out on a very formal basis with pilots quietly eating while discussing some of the test programs they were flying. Then one pilot would throw his dinner roll at another pilot. Other pilots quickly responded, also throwing their rolls. Soon some dipped their rolls in water and pitched them. The dinner typically ended as a major food fight.

The previous Christmas I had attended the party for the first time. During the festivities, someone used scissors to cut off my best tie. Other pilots lost coat sleeves. Because alcohol flowed freely at this dinner, the Air Force scheduled a military bus and driver to transport us from Edwards to Burbank and back again to Edwards. It was about a two-hour bus ride each way.

The party at Lockheed ended around two in the morning, and on the drive back to Edwards one of the tires on the bus went flat. At this point we were anxious to get home.

Some pilots chanted, "Go on, go on, go on," since we weren't responsible for the status of the bus.

Wisely the driver did not take our advice and halted the bus. Around three in the morning the driver jacked up the bus and put on a spare tire. Several of the pilots who had been sleeping through the return trip, woke up to find themselves in the center of the Mojave Desert in the middle of the night wondering where they were and what had happened.

Bob Gilliland's test plane was located in Lockheed's hangar at Air Force Plant 42 at Palmdale, California. The SR-71 pilot and chase pilot would have to rendezvous in the air. I waited for a phone call from Gilliland to set up the details of our mission.

"How are you guys doing?" Gilliland asked. "Recover from our party?"

"Sure," I answered, "but transportation back and forth to Burbank in a Blackbird would have been a lot nicer than our bus."

Bob Gilliland got down to business, outlined his mission, and briefly sketched the flight plan. He would make an afterburner takeoff from Palmdale followed by a turn north to climb over Rosamond Dry Lake and then level off around the town of Mojave.

After I joined up he would shut down one engine and the other would be placed in full afterburner power. He would reduce air speed slowly to determine the minimum flying speed on one engine under heavyweight conditions. It was a hazardous flight and would take a great amount of skill to perform accurately.

After the single-engine test, Gilliland would be performing a Mach 3 speed run at high altitude. He would be wearing a full-pressure suit, which would restrict his mobility and visibility in the cockpit for performing the single-engine test.

"Suppose we do this," I suggested, looking at my watch. "I'll take off five minutes before you and climb over Rosamond Dry Lake. I'll switch to Palmdale tower on the radio and watch the runway to see you roll."

Gilliland agreed, adding, "Now if you miss me after takeoff, I'll be in a climbing right turn and . . ."

"Don't worry," I interrupted. "I'll be there."

Seeing the Blackbird in flight was a wondrous sight, one not to be missed. Gilliland's flight would be canceled if I couldn't join with him, and I didn't plan to let him down. We synchronized time on our watches and were on our way.

I eased the throttle of the Starfighter forward as the sleek interceptor started its take-off roll, slowly at first then faster as the powerful General Electric J79 sucked in increasing quantities of air. The plane lifted off and I climbed to 5,000 feet over Rosamond Dry Lake. As Gilliland started his take-off roll westbound down the Palmdale runway, I eased the nose of the F-104A over and started to pick up speed. When he got airborne I increased power and continued to descend. Gilliland banked the SR-71 to the right. I was in his four o'clock position about a half a mile behind him and closing fast.

Within seconds I called, "Dutch 51, the Air Force is at your service."

"I expected nothing less," he replied.

Over the town of Mojave, fifteen miles northwest of Edwards, Gilliland started his single-engine test. Soon the long sleek nose of the SR-71 was canted up, and the plane was flying at low speed and a high angle of attack. Gradually Gilliland applied full rudder to counteract the twisting motion of the good engine, which was producing thousands of pounds of thrust. He was also applying elevon deflection to keep the wings of the Blackbird level. This once-sleek streamlined SR-71 looked twisted and contorted, as it flew slower and slower, not at the high Mach number for which it would become so famous in the future.

When Gilliland completed his single-engine tests, he restarted the Blackbird's engine. He increased his thrust to full afterburner power on both engines and started to climb and accelerate northbound, paralleling the Sierra Nevada range. My Starfighter could keep up with him in the subsonic speed range. It was a beautiful sight looking down at the Blackbird with the snow-capped Sierras in the background.

Bob Gilliland had a date with a tanker farther north around Fallon, Nevada, and I was climbing and accelerating with him, using lots of fuel, my home base miles behind me. It was time for me to break off and return to Edwards.

"So long, Dutch 51," I transmitted on the radio. "Chase breaking off."

"Good-bye," Bob replied. "You can host the next party."

During my year at the test pilot school everyone knew that Colonel Yeager was part of the old school way of thinking about flight test. Yeager held that all a test pilot needed was that special touch or feel to fly and evaluate airplanes. As students we were part of the new school approach and received extensive academic instruction in evaluating both the performance and the stability and control of military aircraft. We flew test flights and collected data that we evaluated in the classroom. From the data we determined whether the aircraft met military specifications. A special touch or feel was secondary to the data. Flight instruction had been given to us in some of the frontline aircraft of that time, the B-57, T-38, F-101, F-104, and F-106.

Dave Livingston, twenty-nine, one of my classmates from the school, was also transferred to the Fighter Branch of Test Operations. Livingston, a 1958 graduate of the U.S. Military Academy at West Point, had flown the F-105 Thunderchief in Europe before attending the school. He was single and living in the bachelor officers quarters (BOQ).

Livingston and I were captains and considered ourselves to be good stick-and-rudder fighter pilots. We were trained by the test pilot school to expertly handle any fighter manufactured in the United States, even if it was a turkey or loser and difficult to fly. After years of preparation for the best job in the Air Force, we were now performing the primary mission of the Fighter Branch, evaluating new fighters such as the F-4 Phantom, F-5 Freedom Fighter, and F-111A Aardvark.

While assigned to the Fighter Branch Livingston heard about a new sport vehicle called the desert Sand Sailor. Actually, Sand Sailor was a misnomer. The desert lake bed was made up of packed silt and clay. But no one would call the vehicle a silt sailor or clay sailor!

One Saturday morning we drove about twenty-five miles southeast of Edwards to El Mirage Dry Lake, where we met with the designer, John Schindler, who also built and sold his Sand Sailors.

The Sand Sailor was a fifty-pound, three-wheeled vehicle with a fourteen-foot mast and triangular nylon sail. With room for only one person, the machine took advantage of the strong Mojave Desert winds and perfectly flat lake bed to move at up to two-and-a-half times the velocity of the wind.

You belted yourself into a plastic bucket seat located only a foot above the desert floor, just forward of the rear axle.

Connected to the base of the mast was a rudder bar with direct linkage cables going forward to a steerable rubber nose wheel. The Sand Sailor was steered to the left by pushing forward on the right pedal with a foot, or to the right by pushing forward on the left pedal. As pilots, Livingston and I were amazed that the pedals were rigged in that manner. When pilots push the right rudder pedal when taxiing, they want to go right. When they push the left rudder pedal, they want to go left. We looked at each other with disbelief and shook our heads. In our estimation, the builder had made a major design mistake.

Because the winds were blowing only a couple of knots, we each took an introductory ride and were immediately hooked on sand sailing (or land sailing, as the sport is more generally called). We bought a brightly painted yellow model with the number seventy-four sewed on the sail. Then we loaded it on top of my 1960 Rambler station wagon and drove back to Edwards. Since I was married with two small children, I lived in base housing. There was a single-car garage attached to my house, and Livingston and I hung the Sand Sailor in my garage above the station wagon using a system of ropes and pulleys.

Rogers Dry Lake was the primary emergency landing lake bed for disabled Air Force aircraft and the standard landing field for the X-15 rocketship. Military authorities at Edwards would not let us sand sail on Rogers, the main lake bed adjoining the base. They thought that we might interfere with pilots using it for emergency landings.

We applied for and were approved to use Rosamond Dry Lake, located about four miles west of the base. The approval required us to make a phone call to the Edwards control tower in advance of each use to ask for permission. Pilots of test aircraft would then be advised to watch out for us since we were without radio communication with the tower.

Operating the Sand Sailor required both steering with your feet and controlling the sail with your hands. A five-foot rope was tied to the base of the sail, which was then threaded through a pulley on a metal bar about two feet behind the seat. We held the end of the rope in both hands so that we could adjust the angle of the sail in relation to the direction of the wind.

To obtain maximum speed, we steered about 45 degrees off the direction of the surface wind and held the rope taut. Once we got up to speed, we could make a hard 90-degree turn crossing the tailwind vector and pick up speed in another direction. We could also tack upwind by making a series of zigzag courses in order to return to our starting point. Similarly to snow skiing, it took a couple of turns to feel out the steering and proper angle of the sail.

In only a few minutes we felt comfortable and were into the joy of sailing without water.

Land sailing could be dangerous if a person tipped over at high speed. If a gust of wind or improper steering started to lift the sailor up on just two wheels, the pilot simply let go of the rope, brought his or her hands toward the stomach, and let the Sand Sailor overturn. The fourteen-foot mast acted like a roll bar, and you wouldn't get hurt if you used proper procedures. When the dust settled, you unhooked your seat belt, fell to the lake bed, got back up, and tried it again.

Before long, Livingston and I could both maneuver the Sand Sailor like true fighter pilots, doing figure eights around the lake bed. Sitting so low to the ground we had a tremendous feeling of speed, but we wondered how fast we were actually going. A couple of rocks and bottles were placed at a measured distance apart and we tried to time the seconds it took to go from one to the other. Figuring our speed that way was an involved process and took too much time. Fighter pilots are not known for patience, so we needed a better system to measure speed.

In flight test, a new fighter's airspeed system is calibrated by flying in stabilized close formation with an already calibrated aircraft and then comparing airspeed readings. So, as Livingston maneuvered the Sand Sailor, I drove in formation in my station wagon and attempted to read my speedometer. We made a couple of data passes, but each time the test degraded into rat-racing, a playful form of high-speed flying in which airplanes pursue or attempt to outmaneuver each other, similar to a game of cat and mouse. Both of us were trying to get into the six-o'clock position of the other for a gun pass. The Sand Sailor could turn extremely tight into the wind and the station wagon would overshoot. Turning downwind, the Sand Sailor would accelerate fast and come close to tipping over. Here the station wagon would gain an advantage. We knew that the station wagon would win if we collided.

As we experimented with what test pilots would call the handling characteristics of the Sand Sailor, both of us became very proficient at sensing the proper wind angle at which to steer the Sand Sailor and the amount of force required on the rope to obtain a maximum speed of about 55 knots. We had developed that Yeager touch or feel. Now we lived for the day when Edwards test flying was canceled due to strong surface winds. That was the opportunity to go out to Rosamond Dry Lake in search of what in surfing was called the perfect wave—for us land sailors, the perfect wind. When the wind was so strong that grains of sand stung our faces, we felt like open-cockpit aviators in the early days of aviation. It was an exquisite moment listening to the sail fluttering and snapping while flying like the wind in the Sand Sailor.

Soon we spent all our spare time thinking about ways to improve the speed performance of our Sand Sailor. We talked to aircraft test engineers with whom we worked on our test programs. What could we do to gain a few more knots of speed? They recommended we install an airspeed system or an angle-of-attack vane. The engineers were also part of the new school form of testing—all you needed was more recording instruments and data to evaluate.

Then we recalled the day we purchased the Sand Sailor and had questioned designer John Schindler about his backward steering arrangement. Maybe we could alter the steering and obtain a few more precious knots. We noticed that the metal cables that connected the rudder bar to the nose-wheel axle were simply attached with cotter keys. It dawned on us that if we disconnected the cables and reversed the attachment at the nose-wheel axle, we could configure the steering like an airplane. In an aircraft on the ground you push right foot forward and turn right. You push left foot forward and turn left.

In a couple of minutes we were ready for a test run. To our astonishment, the Sand Sailor was almost uncontrollable. On straightaway speed runs we could handle it, but for tight maneuvering both of us just about lost control. How could that be? The only answer we could think of at first was that we had become so familiar with the original configuration that we had adapted to the builder's poor design.

Another explanation concerned watching the nose wheel during a turn. In a tricycle landing gear plane you cannot see your nose gear from the cockpit. Likewise, in a conventional landing gear airplane, you cannot see the tail wheel. From our sitting position about four feet behind the twelve-inch diameter nose tire the wheel was a powerful visual image. Maybe we were unconsciously looking at the tire during turns.

We thought of a method to solve our steering dilemma. Our plan was to invite another test pilot from the Fighter Branch to evaluate the Sand Sailor without knowing the original design. Capt. Lawrence C. "Larry" Curtis Jr., like Livingston, was a former F-105 pilot from Europe who had just finished the test pilot school and was now assigned to our Fighter Branch. Curtis, thirty, from Santa Clara, California, was very tall and thin and had dark black hair. He was married with three small daughters and also lived in base housing.

Curtis got his introductory ride with the steering cables crossed. He did not know we had altered the original design. Just as we had experienced, Curtis found the Sand Sailor extremely difficult to steer. Familiarity was not the culprit.

Curtis, who also thought of himself as a good stick-and-rudder pilot, was quite embarrassed that he could not handle the Sand Sailor better. As test pilots, we all prided ourselves in having good hand-eye coordination and the ability to handle any vehicle on the ground or in the air. Livingston and I explained to Curtis how we had crossed the cables and, as soon as we returned the steering to the original configuration, he did fine.

At that point, we needed to reevaluate our thoughts on the man-machine interface of the Sand Sailor, a technique we were taught in test pilot school. Why was the steering in both an aircraft and the Sand Sailor correct—but opposite one from the other? We thought about other vehicles that are designed to steer on the ground—bicycles, tricycles, motorcycles, snow sleds, scooters, and automobiles. All steered exactly like the Sand Sailor, whether a person used their hands or feet. All steering was natural and easy. Obviously the Sand Sailor designer was correct after all.

Then we wondered whether it was the airplane that was designed wrong. Because the earliest planes were of a conventional landing gear design (tail-wheel aircraft), turns left and right were accomplished by using rudder pedal movement and engine power to provide airflow over the vertical tail. Just like the Sand Sailor, the easiest way to connect the rudder pedals to the aircraft rudder was direct cable attachment. But this connection was to the rear rather than the front as in the Sand Sailor. In the aircraft, moving right rudder pedal forward pulled the aircraft rudder to the right and the aircraft turned right. Moving the left rudder pedal forward pulled the aircraft rudder to the left and the aircraft turned left. In the air the rudder was used to coordinate a banking turn and was rigged properly.

Flying must have been such a unique experience for the pioneer aviators who accepted this flight-control rigging and never questioned it, even though they had previously driven bicycles and other ground vehicles. Once pilots learned to fly a particular way, every aircraft designer after that had to follow the same scheme.

I began to wonder about the lessons learned from our land-sailing experiences. With our technical aviation background, if Livingston and I had designed the first land yacht we would have used the aircraft method of steering. That method would have been an error on our part.

Something else concerned me since my first speed run on Rosamond Dry Lake. If I experienced an in-flight emergency in one of our experimental jet fighters and had to land on the lake bed, would my mind revert to my Sand Sailor steering techniques? I then might apply the wrong rudder movement, lose control of the aircraft, and crash.

In my previous years of flying, I didn't have to think about moving the flight-control stick or rudders. Correct stick and rudder movements seemed natural for the aircraft motion I desired. Some fighters had stiff or heavy controls, others weak or light. After a few training flights, I adjusted to the feel and didn't think about it again. But like playing Ping-Pong, sometimes it seemed that the hand moved faster than the brain. Now my mind had been altered by the thrill and joy of flying like the wind in the Sand Sailor. Some part of my brain was programmed for flying and another part for land sailing.

As I look back, although technology has greatly changed the way aircraft are tested, it is clear that a test pilot's touch or feel will always be as important as brain power. It would have been interesting to invite Yeager to take a test run on the Sand Sailor with the cables crossed.

Maj. Tommie D. "Doug" Benefield was a member of the test pilot staff when I arrived at Test Operations. His specialty was bombers and heavy transports, although he had checked out as first pilot in over sixty types of aircraft. In his career he had flown fighters, cargo transports, personnel carriers, bombers, light planes, and a variety of experimental types, both piston and jet powered. Benefield even tried his hand at helicopters. It was his contention that a good test pilot ought to be able to fly anything. He lost no opportunity to check out in whatever showed up on the flight line that was new and different and to which he had access.

Doug Benefield, thirty-six, from Rison, Arkansas, was a large and vigorous man. Taller than six feet two inches, he weighed well over 240 pounds. In his bulky flying suit his physical presence was considerable. He exuded an aura that suggested that he was immediately ready to fly whatever complicated new jet transport awaited or to hit the middle of the line of the University of Southern California football team for six yards and a first down.

Benefield had an air of relaxed confidence that was almost an essential qualification in our exacting profession. He spoke easily on the subject of flying, which, as far as he was concerned, was inexhaustible. Like the other guys I knew on the test pilot staff, he clearly loved his job and refused to trade it for any other job in the military service. Benefield had received several civilian test pilot job offers, but he preferred the Air Force with its constant challenges. Like Frank Liethen, Benefield usually relaxed with a big cigar, secure in the knowledge that he had thousands of hours tucked away in his logbook and a top-notch job.

Benefield had found his specialty—heavy aircraft—as a first lieutenant. Early in his career he became a pilot on the Douglas C-124 Globemaster, a giant of a plane that allowed huge pieces of equipment to enter through its massive clamshell doors and then could fly them to distant parts of the world. A need arose for nine Globemasters for an unusual and record-setting mission. Benefield and his crew were among those picked. With eight other planes they flew to Paris, France, loaded paratroopers, and took them to Indochina. When the airlift was completed, the big planes continued on around the world and arrived home again, all in the space of twelve days.

When the Korean War erupted Doug Benefield was promptly shipped to the scene of action. He flew 328 combat hours over the rugged terrain and got his baptism by fire, putting to hard use the lessons from flight school and those that he had learned from experience.

While overseas he took part in one of the most famous airlifts in history. Maj. Dean Hess, another combat pilot, had gathered a small army of almost nine hundred homeless Korean children whose town was about to be overrun by a Communist advance. Hess gathered them together at an airfield from which they were flown to safety. The story was later told in Colonel Hess's book *Battle Hymn*, which was also transferred to the movie screen. Doug Benefield was one of the pilots who loaded up many of the otherwise condemned children (they could have been killed) and flew them away to a new orphanage.

Shortly after returning home, Benefield drew an assignment that was to have much to do with his future career. It came in the unsuspected form of a goodwill tour of South America. The Thunderbirds, the famous Air Force acrobatic demonstration team, were part of the tour. Benefield flew the transport that was loaded with extra engines and spare parts for the exhibition aircraft. This interesting but still relatively routine assignment gained added significance because an Air Force test pilot was going along—Chuck Yeager. During the tour Benefield had an opportunity to talk with Yeager. After Benefield had done so, he knew he wanted to be a test pilot too.

Benefield was selected to join the 1955 class of the test pilot school starting in January 1955. For eight months he studied, flew, "reduced" data (plotting test points on graph paper), and wrote lengthy reports. Among those who looked over the graduating student test pilots was Lt. Col. Frank "Pete" Everest, one of the Air Force's rocket test pilots. He saw in Doug Benefield the kind of man he wanted for the test pilot staff that he commanded. The boy who had once built model planes and spent long hours pondering the mysteries of flight had finally become the man who had made it to the best place on earth to test heavy aircraft: the Bomber/Transport Branch of Test Operations.

Flight Test Operations was the big league of flight testing or "the Show" as they say in major league baseball. I couldn't miss Benefield when I first arrived at Test Operations. He had the type of personality you immediately liked. He was a robust man, always with that cigar in his mouth, and looked like he would be a tight fit in a fighter aircraft. Benefield was a major and six years older than I.

As a junior test pilot I got a call from senior test pilot Benefield. "I'm going to be doing some testing in the Douglas C-133 Cargomaster transport tomorrow," he said. "Would a fighter pilot like to come along?"

"Yes," I said, never one to turn down a flight.

The C-133 was a high-wing, very heavy cargo/transport aircraft. The plane weighed 120,000 pounds empty and 286,000 pounds at maximum gross weight. This was about three to seven times the weight of any fighter aircraft I had ever flown. The C-133 had a wingspan of 180 feet and was powered by four Pratt & Whitney turboprop T34-P-3 engines, making it the largest turboprop aircraft in military service.

The Air Force had experienced several recent mysterious accidents in the aircraft in squadron use. Benefield was part of the test team formed to come up with suspected cause factors, make up a flight test plan, conduct the flight tests, and then reduce the data. And, finally, he would pass the conclusions and recommendations on to the Air Force higher command.

The next year I was assigned an F-4C Phantom stability and control test program to investigate the cause of several accidents in that aircraft. My flying experience and discussions with Benefield gave me a head start and prepared me to evaluate the flight controls of a fighter aircraft.

To a fighter pilot, the first impression you get flying large transport aircraft is the feel of heavy flight-control forces. All fighter aircraft are flown with the right hand on the control stick and the left hand on the throttle. The aileron, elevator, and rudder forces are sports-car light in a fighter-type aircraft so that the pilot can easily maneuver. Even pulling 6 Gs takes only about twenty pounds of force.

As I sat in the copilot's seat of the C-133, I had my left hand on the throttles and my right hand on the yoke. To make a turn, I placed both hands on the yoke, rotated the yoke clockwise for a right bank or counterclockwise for a left bank. The control forces were very heavy because transport aircraft were designed for only a couple of Gs, and the manufacturer didn't want the pilot to be able to overstress the aircraft.

After I got comfortable flying the aircraft, Benefield said, "Let me have it a moment."

He took over control of the aircraft, and I looked at him.

"Watch this," he said.

I saw him rapidly rotate the yoke perhaps three quarters of its throw to the right, followed by the same motion to the left. Nothing happened; I was not impressed.

"So what?" I started to say when an earthquake type of motion wildly shook the cockpit sideways for quite a few seconds.

Any crew member standing up would have been immediately thrown to the floor. Those of us strapped in with a seat belt and shoulder harness were secure from injury, but the severe motion and groaning sound of the aircraft structure was breathtaking. Obviously this was not a good characteristic in an aircraft. Possibly even air turbulence could excite this strange motion.

Benefield then explained to me, "The rapid rotation of the control yoke caused the ailerons to excite the natural structural frequency of the outer wing, causing a vibration that started at the tip of the wing and advanced toward the fuselage. When the vibration contacted the fuselage, it migrated forward to the cockpit and backward to the tail."

"Wow," I said, "that shaking can't do the aircraft structure any good."

Benefield replied, "I think this bad characteristic is the reason for some of the C-133 crashes. I thought you might find this flight experience of value in your future testing. I've never flown any transport aircraft that responds so poorly to control inputs."

We finished the test flight, and I made two touch-and-goes followed by a full-stop landing.

This fighter pilot walked away with great respect and admiration for Doug Benefield. I was impressed by how cautiously he handled the flight tests, how organized he ran the crew members, and how well he handled the added burden of responsibility that comes in a very large aircraft. As a result of the flight tests Benefield performed, the size of all of the C-133's dorsal fins was drastically increased to greatly improve the stability of the aircraft. My flight with him was an unforgettable experience.

After Dave Livingston and I had been land sailing for some months I was scheduled to chase Northrop test pilot Richard "Dick" Thomas as he flew a test mission in the Northrop F-5A. Thomas, thirty-five, had earned his Air Force pilot wings in an open-cockpit Stearman in 1952. He flew the F-80C, T-33, F-86, and F-100D aircraft for six years before leaving the military for a flight test career with Beech Aircraft and later Boeing. Thomas tested the RB-47 and B-52 for Boeing; in 1962 Boeing sent him to the U.S. Naval Test Pilot School. He joined Northrop Aircraft in 1963.

An Air Force photographer was to accompany me in the backseat of my T-38 Talon and photograph the F-5A as Thomas dropped several bombs in the Edwards Precision Impact Range Area (PIRA). Prior to flight, Thomas called me on the phone and described the flight path and speed he would use on his weapons delivery pass. My slower T-38 would be able to keep up with him because his bomb load would slow his plane. The entire test mission would only take about twenty minutes, but, unknown to me at the time, it would bring back my land-sailing skills.

We joined up on the taxiway leading to the Edwards main runway. Thomas took off and I waited a few moments on the runway before I rolled.

Within a couple of minutes Edwards approach control vectored us over the bomb range, and Thomas dropped his bombs. My photographer got the required photos, and we broke off formation. Dick Thomas returned to land at Edwards.

With over two hours of fuel remaining, I decided to practice touch-and-go landings at Palmdale. Just after my second landing at Palmdale, the photographer and I heard a strange sound in the Talon. I checked the flight and engine instruments—everything looked normal. However, we both could smell fuel fumes and decided to select 100 percent oxygen in our facemasks.

I turned the T-38 toward Edwards, started to climb, and radioed Palmdale control tower that we were heading home. Rechecking my flight instruments, I noticed that the pointers on both the left and right fuel gauges were moving downward. The twin GE J85 engines would not normally consume fuel that fast, so I thought I must have a massive fuel leak.

Climbing through 6,000 feet, I switched the ARC-34 UHF radio to Edwards tower frequency. Unfortunately, the radio did not lock on to the frequency but started to cycle through all the preset channels. I selected manual, but the radio continued to "click-click-click" in my helmet. I turned the radio to "Off," waited thirty seconds, and turned it on again. "Click-click-click," it continued.

Without radio contact with Edwards tower, I was unable to declare an emergency and request priority handling into the Edwards pattern. Passing 10,000 feet directly above the Edwards runway, both low-level fuel warning lights illuminated. Total fuel starvation was now only seconds away. If both engines quit, I would lose aircraft electrical power and hydraulics for the landing gear, flaps, and speed brakes. Even more important, I would lose hydraulic power for the ailerons, stabilizer, and rudder.

Maj. Dave Tittle, my Talon instructor in test pilot school, had cautioned me earlier that a dead-stick landing in the T-38 was impossible to accomplish. The only option I had, according to him, was to eject. Even though it was a bright and beautiful day at Edwards, I was in a very difficult and dangerous situation.

Then a thought came to my mind. If I lowered the landing gear and selected half flaps while the engines were still running, maybe I could dead stick the T-38 onto Rogers Dry Lake. The wind was calm—I could land in any direction I chose. I knew Livingston was not on the lake bed with our Sand Sailor and I wouldn't have to avoid him. There were several pools of water on the lake bed that remained from the winter rains, but I knew I could avoid them and select a dry area to land.

To maintain hydraulic power for the flight controls, I could dive the T-38 at the maximum gear-down speed of 250 knots—hoping that engine-windmilling speed would keep the flight-control pressure up. If I leveled off ten feet above the lake bed and let the airspeed slowly bleed off; maybe I could land and save this beautiful bird.

I lowered the gear and half flaps and informed the photographer on the intercom to "hang on."

Without the ability to notify Edwards tower that I was making an emergency landing on the lake bed, we would be all alone. If we crashed it would take a while for someone to notice smoke billowing up from the lake bed and notify the fire department.

With the nose pointed down 20 degrees and both engines in idle, I aimed for the center of the lake bed parallel to the direction where the X-15 rocketship landed after a test mission. Suddenly, I heard the engines flame out and saw red warning lights in my peripheral vision. I was tempted to look at the engine RPM but resisted doing so, knowing full well the RPM was going down not up. Seeing the fuel gauges at zero wouldn't improve my mental outlook either.

As planned, I dove toward the lake bed using the minimum amount of aileron and stabilizer movement. The Rogers lake bed was fast approaching as I gradually pulled the stick slightly aft. Little by little the rate of descent reduced, and I raced above the ground now doing 200 knots. I brought the nose up a little more and settled just above the desert lake bed, still a couple of feet in the air. I touched down at around 140 knots and saw miles and miles of lake bed still in front of me. The nose of the Talon kissed the ground and I gently touched the manual brakes. Gradually we slowed down and finally came to a stop. Even though I had landed on the lake bed, I flew the T-38 like an airplane, not a Sand Sailor!

It was then that I became conscious that the UHF radio was still going "click-click-click." I turned the radio off, expecting the cockpit to be completely quiet without the engines running. All I heard then was both of us breathing hard on the intercom.

In preparation for flying the F-111A, my land-sailing friend Dave Livingston and I were scheduled to fly a tandem-seated T-33 to the General Dynamics plant in Fort Worth, Texas, to attend a meeting. The flight to Texas was uneventful. I flew the front seat on the first leg to Davis-Monthan Air Force Base, Arizona, and Dave flew the front seat on the flight to Texas.

We attended the meeting in our military dress uniforms and then hurried out to the flight line to prepare for the flight back to Edwards. I quickly changed out of my dress uniform and put on a flying suit.

On the belly of the T-33 we carried a travel pod. The travel pod was a four-foot-long, three-foot-wide metal container attached to the bottom of the fuselage that held about fifty pounds of luggage. Livingston placed his hang-up bag in the travel pod as I preflighted the T-33. I pushed his bag to the rear of the pod and placed my new nylon hang-up bag in the front portion of the pod. I took my brightly shined black oxford shoes and carefully placed them in the pod in a position where they would not be scratched. Under them I slid a manila folder full of the semiclassified, handwritten notes I had written on the F-111A during the meeting. I then put on my flight boots, closed and locked the pod, climbed the entrance ladder, and got into the front seat.

It was a gorgeous day as we flew at 35,000 feet over west Texas with a planned refueling stop at Kirtland Air Force Base just south of Albuquerque, New Mexico. From about one hundred miles east of Kirtland I could see the base so I started a gradual descent with reduced power.

We came whistling down, passing over the mountains east of the base at 350 knots. As we approached Kirtland at 1,500 feet above the ground, the control tower cleared me for an overhead pattern and directed me to make a break (a 180-degree turn) at midfield. I followed their command and rolled into a 60-degree left bank and dropped the speed brakes to slow the T-33 down to a speed where the landing gear could be extended. Livingston and I both heard a strange noise as the speed brakes opened, an abnormal sound for the T-33. As the wings were leveled on downwind, I checked the hydraulic pressure and engine instruments. All readings were normal and the aircraft seemed to fly satisfactorily. The landing gear and wing flaps were lowered and I made the final turn to line up with the runway. Landing was made on the centerline and I applied light braking.

Suddenly, I noticed one black shoe on the runway and swerved slightly to avoid it. From the back seat Livingston could not see the shoe.

"You are not going to believe this," I said over the intercom. "I just about ran over a black shoe."

As we slowed down, I noticed pieces of paper falling from the sky directly in front of me. The paper looked like leaves twisting and turning in the light wind as they slowly dropped on the runway.

"There must have been a midair collision above the field," I remarked to Livingston. "Where else would all this debris be coming from?"

We taxied off the runway, and an Air Force "follow me" truck met us. I followed the truck and parked in front of base operations. I shut down the jet engine and immediately climbed out of the cockpit. Both Livingston and I inspected the T-33 to see if we could find anything that might have caused the strange noise we had heard.

It was simple enough to find what was wrong—the nose of the travel pod was missing. My new heavy-duty vinyl hang-up bag was gone along with both of my shoes and the manila folder holding the F-111A information. At least I knew where one of my shoes could be found. But most worrisome was the potential loss of the semiclassified data.

Observing our situation, the Kirtland Air Force Base airdrome officer (AO) drove up in his military jeep. AO duty was a twenty-four-hour assignment rotated among all the pilots stationed at the base. It was duty every pilot got about twice a year. The AO reported to the base operations officer and was responsible for signing flight clearances, greeting visiting pilots, monitoring the control tower and weather office, and anything else that affected the operation and safety of the airfield.

What a welcome sight for us to see the AO who happened to be Capt. Jim Hurt. Hurt was a former classmate of ours at the test pilot school, the person who always had a grin on his face. After graduation Hurt was assigned to the Air Force Special Weapons Center at Kirtland where he flew the F-105 Thunderchief and developed new tactics and procedures for the launch of nuclear weapons. Seeing the front of our travel pod missing, Hurt did not have his usual big grin. He was concerned that the objects we lost off the T-33 could damage other aircraft using the Kirtland runway. Livingston went into base operations to close our flight plan and fill out the Air Force forms concerning the loss of an object during flight.

I jumped into the right seat of Hurt's jeep and he drove to the runway. I recovered my one black shoe and several sheets of paper near the centerline of the runway. Surprisingly, the recovered shoe was in perfect condition even though it had fallen from a height of 1,500 feet. Somehow it must have landed on the sole and bounced down the runway without receiving even a small scratch.

Little by little I gathered papers containing the F-111A information. As Hurt drove on the edge of the runway, I ran out to retrieve pieces of crumpled paper as other military aircraft took off and landed. To those pilots, I must have seemed like a humorous figure dashing and darting around the runway chasing elusive sheets of paper blowing in the New Mexico wind. Gradually I picked up most of the paper—the pages were not numbered but with two hands full it seemed like I had enough.

A pilot flying in and out of a military base doesn't realize how large an area is taken up by a runway and surrounding taxiways and ramp. What seemed small from the air was in reality a huge complex covering many, many acres of land. We had an enormous amount of land in which to hunt for rather small objects.

Still missing was my new nylon hang-up bag and my other black shoe. (Livingstone's hang-up bag was in the back of the travel pod, held in by air pressure.) Before a travel pod was designed for the T-33, hang-up bags were stored in the nose of the plane next to the battery. The first couple of hang-up bags I had were commercial bags made of very thin vinyl and clear in color. They did not last very long being exposed to battery acid and the normal grease and grime associated with aircraft.

Seeing me arrive on a cross-country flight with a bag torn and tattered, my mother purchased a heavy-duty blue nylon bag for me that was guaranteed to withstand the rigors of flight. It was now hidden somewhere within three square miles of New Mexico runway and sagebrush. My nylon bag had taken a flying leap from the aircraft and flown 1,500 feet to an unknown landing. About fifteen minutes later we found the bag draped over a bush, looking like it had carefully been placed there for safekeeping. It was in perfect condition—not even any dust or dirt on it.

Searching for the remaining black shoe proved to be a challenge. We drove back and forth, around and around. After a while we weren't sure where we had searched and where we had not. It was like looking for a needle in a haystack. If both shoes had been missing, I would have given up quickly and forgot about them. But I had one mint condition shoe that had spent time on a runway narrowly being missed by aircraft taking off and landing. The still missing shoe must also have experienced an unusual flight and crash landing.

We continued to hunt as the sun was close to setting in the western New Mexico sky. The shadows became longer and hope was fading that the shoe would ever be found. With the last bit of sunshine on my back and looking eastward, I thought I saw a dark object under a bush some distance away. I pointed in the direction of the potential shoe location, but Hurt could not see it. Concerned that the speck of black would disappear if I took my eyes off of it, I gave Hurt directions to drive to the area. There, hidden under a bush, was the other shoe, scratched, scraped, and full of dirt.

Somehow the shoe that had fallen on the concrete runway had come away unscathed, but the shoe that had fallen into the New Mexican brush was severely damaged. As much as I tried to polish out the marks, I could never get the shoe to shine brightly again.

In the days to follow, I wore these dress shoes to many military formations and even dances at the Officers Club. When they eventually wore out and were discarded, I felt sorrowful. Whenever I wore those shoes, I felt lightness in my step and remembered that my flying shoes had taken a flight I as a pilot had never experienced.

The first of the variable-sweep wing General Dynamics (GD) F-111As arrived at Edwards in the middle of January 1966. The plane was serial number

63-9773, a preproduction aircraft and the eighth plane built for the Air Force. Test pilots would subject the F-111A to simulated combat situations under day and night and all-weather conditions. The flight tests would include aircraft performance, stability and control, evaluation of the weapon system performance, and reliability and maintainability with emphasis on the bombing, terrain-following radar, communications, penetration aids, and engine propulsion.

Lt. Col. James W. "Woody" Wood and Maj. Robert K. "Bob" Parsons, the first two Air Force pilots checked out in the F-111A, were at the controls as the fighter-bomber made its flight from the GD manufacturing and flight test facility at Carswell Air Force Base, Texas, to the Air Force Flight Test Center. Colonel Wood, forty-two, from Paragould, Arkansas, had learned to fly in 1944 in the Army Air Forces. He flew combat for a year from England over Belgium and Germany. After World War II, Wood attended college and gave flight instruction in civilian schools. He was recalled to active duty in 1948 and served in the Far East Air Force. Wood graduated from the test pilot school in 1957 and tested all the Century Series class of fighters. He was selected as a pilot for the Dyna-Soar project but returned to Fighter Operations after the program was canceled. Major Parsons, thirty-six, was born in Ripley, Virginia. He attended the test pilot school in 1962 and applied to become a NASA astronaut. Parsons was turned down by NASA but was selected for the Aerospace Research Pilot School Class IV. Wood and Parsons would both test the F-111A and train other test pilots to fly the aircraft.

Because of its size and design, the F-111A was really more of a bomber than a fighter—and perhaps the most controversial weapon since gunpowder. General Dynamics said it was the best airplane in history. Senator John L. McClellan, chairman of both the Senate Appropriations Committee and the Committee of Government Operations, called it a flop, a multibillion-dollar blunder. The F-111, originally a paper plane (design only) called the TFX (for tactical fighter, experimental), was received with so much enthusiasm by the civilians who ran the Pentagon that they decided all three services should have it. The Army begged off early, but the Air Force and Navy were forced by Secretary of Defense Robert S. McNamara into a mixed marriage that neither really wanted. That was only part of the controversy. The fight for the $10 billion contract was spectacular, and, when Boeing (thought by many to have the best design) didn't get it, screams of politics were cosmic. All of which pointed up the fact that no other plane in history was born under such a cloud of suspicion. Secretary McNamara wanted a plane that used commonality between the parts of both the Air Force and Navy models. He rammed the concept through despite the opposition of practically everyone in sight. It was a colossal mistake.

Another problem soon surfaced as costs grew tremendously. In 1963, McNamara told Congress the F-111 would cost about $3 million a copy. But it eventually exceeded more than twice that amount. Originally, commonality was supposed to save a billion dollars. But most experts agree that solving the problems brought on by commonality ate up those billion dollars and perhaps much more.

From a size perspective the F-111A was an impressive bird. It was 73 feet 6 inches long. Its wingspan was 63 feet with the wings cranked out to 16 degrees and just 31 feet 11 inches with the wings folded back to 72.5 degrees for supersonic flight. The swing-wing brought many important advantages: the plane could fly at two-and-a-half times the speed of sound at 60,000 feet and come roaring in at Mach 1.2 (about 800 mph) at treetop level. Yet the two pilots, sitting side by side, could loiter on target for hours with the wing swept out. The aircraft was to be used primarily for offense as a penetration and interdiction weapon. Almost anything could be hung on the underwing pylon stations: forty-eight 750-pound iron bombs, rockets, or nuclear weapons. Fully loaded and ready to fly, the F-111A weighed upward of 80,000 pounds.

Capt. Dave Livingston, my land-sailing friend, and Capt. Herb Brightwell, a TAC pilot, and I attended F-111A ground school at the General Dynamics facility at Edwards. We spent hours and hours learning about the plane's flight limitations, hydraulics, electrical, pneumatics, and other aircraft systems. The instruction was very comprehensive, and we had volumes of documents to read each evening. During our training, the ninth F-111A, serial number 63-9774, arrived. Livingston, Brightwell, and I would be flying both aircraft. Since each aircraft was a handmade, one-of-a-kind aircraft, the flight limitations and systems were different. Keeping track of the information was a monumental task.

To become more familiar with the F-111A, I flew a T-38 several times to the GD facility in Texas to act as a safety chase. Richard L. "Dick" Johnson was the chief test pilot for GD. Johnson, forty-nine, was born in Cooperstown, North Dakota, and joined the Army Air Forces in 1943. He became a fighter pilot and completed 180 combat missions in the Republic P-47 Thunderbolt. After World War II Johnson became chief of fighter test at Wright Field (now part of Wright-Patterson Air Force Base) in Dayton, Ohio. While temporarily assigned to Edwards, Major Johnson piloted an F-86 Sabrejet to a world speed record of 680 mph during four low-level passes over the lake bed. For that accomplishment, he received the prestigious Thompson Trophy. Johnson joined the Convair Division of General Dynamics in San Diego and made the first flight of the F-102 Delta Dagger and F-106 Delta Dart. On December 21, 1964, Johnson and test pilot Val Prahl made the first flight of the F-111A.

Johnson was described to me as a slim, "grim-faced icicle." I found it to be an accurate description.

Flying as safety chase on the F-111A for Johnson and Prahl was a rare opportunity for me. Not only did I learn the current status of the test program but I also got to see the new bird in the air. On most occasions a GD photographer flew in the backseat of my T-38 and documented the test mission on film.

The previous year I had purchased a small Argus C3 35mm camera and carried it in my helmet bag whenever I flew. I took color slides of other aircraft during test flights but had difficulty because it took two hands to focus the camera, adjust the aperture, and fire a shot. It was very hard to fly formation with my left hand on the twin throttles, my right hand on the control stick, and still take photos. The GD photographer built a metal handgrip for me. This device allowed me to operate the camera with just my left hand. Then I was able to safely take photos and still fly in close formation. After several chase missions I was anxious to start testing the F-111A myself.

In May 1966 I made my first flight in the F-111A with Major Bob Parsons sitting in the right seat. Because of engine-compressor stall problems and the fact that the plane was severely overweight, our Edwards test group was not able to do performance and stability and control tests. The F-111A was being redesigned and our number eight and nine aircraft were not representative of production aircraft. So tests of the weapon system were all that we could do. Over the next month I flew four more flights to qualify in the plane and also test the General Electric air-to-ground radar, Texas Instruments TFR (terrain-following radar), and Litton inertial navigation system.

Flying the F-111A was a real disappointment to me. The cockpit visibility was poor due to the side-by-side seating. The plane was so overweight that it just staggered up to altitude in military power (full throttle without afterburner). The plane could barely reach 25,000 feet, and that was without any armament on board. The engines were slow to accelerate from idle to full power, a real disadvantage for a fighter.

All new aircraft have some birthing problems, but not as severe as the F-111A. The plane was also a maintenance nightmare. Test flight after test flight was canceled, with me sitting in the cockpit on the ramp with the engines running because some system, such as the flaps, the inlet ramps, or nose-wheel steering, had failed. Likewise, half of my airborne test flights were terminated early due to component failures. The test program was falling behind schedule and the pilots were discouraged. A military Teletype message that we received from Secretary McNamara ordered the test pilots "to be optimistic about the F-111."

"What a joke," I exclaimed. "The F-111A is the worst aircraft I have ever flown!"

The next month I was involved with one of the most unusual tests a pilot could conduct at Edwards during summer months in the Mojave Desert: wet runway testing. For three days in the latter part of June, Major Bob Parsons and I flew braking tests in the number nine F-111A. On June 24 we made two flights and performed maximum braking action on a dry runway. This data was used as a baseline for the tests to follow. The following day we made six flights of about fifteen minutes each and used full brakes on a wet runway.

For each flight we took off and orbited until the fire department sprayed water on several thousand feet of the runway. Then we made an approach at a selected airspeed and landed precisely on a marked spot on the runway. As soon as the F-111A touched down we applied maximum braking with the foot brakes while maintaining the plane on the runway centerline with the rudder. Following these exacting procedures required good piloting skills. In addition, there was always the chance some mechanical malfunction could spoil our whole day. We were very vigilant on each flight.

Between each flight the ground crew jacked the aircraft above the taxiway, cooled the brakes, reloaded the wingtip cameras, and installed new cartridges for engine start. The F-111A was also weighed after each flight. The unique tests—made three days ahead of schedule—were for checking the variable-wing landing performance so that data could be inserted in the pilot's flight manual. Six flights in one day for one aircraft was a new record for Edwards. The previous record was three flights in one day. The GD public relations office submitted a press release stating the results were "highly satisfactory." However, regardless of the results, GD always made a positive claim.

On the third day Major Parsons and I flew four more flights to complete the braking tests. We had flown fifteen flights in the F-111A, each at a different weight, and logged just over three hours of flying time. All of the missions were under the extremely hot Mojave sun—a grueling effort—for such little flying time. These tests did not endear me to the new F-111A either.

I agreed with Maj. John Boyd, a Pentagon expert in fighter tactics, when he said that "the only good thing about the F-111 is that the dumb-ass Soviets believed our propaganda and built their very own piece of useless crap, the Backfire bomber. The F-111 should be painted yellow and used as a school bus."

They say a cat has nine lives. How many lives does a bird dog have? I flew many different aircraft for months without experiencing very many serious mechanical malfunctions that could cause an airborne emergency.

Some pilots, though probably due to the luck of the draw, always seemed to get the plane that comes apart in the air.

Maj. Mike Adams, the pilot who checked me out in the F-5A, was one of those pilots. Mike, thirty-five, from Sacramento, California, was a laid-back, easygoing fighter test pilot. He graduated from the test pilot school just as I started my year of training in the school. Unlike most fighter pilots, who are loud and boisterous, Adams was quiet and unassuming, but he was very confident of his flying skills. He liked to hunt in the Mojave Desert and had a shooter's eyes.

While I attended test pilot school in 1964, Adams was at Test Operations conducting stability and control tests on the F-5, the fighter version of the T-38. He was perfect for that assignment, having previously served as a fighter-bomber pilot during the Korean War. During his four months of service in Korea that ended with the signing of the Panmunjeon peace pact, Adams flew forty-nine combat missions and was awarded the Air Medal.

As a student in the test pilot school, Adams made a split-second decision that saved his life during a landing accident in an F-104. Adams was riding the backseat in a tandem-seated F-104B piloted by a fellow student, Capt. Dave Scott (later selected by NASA as an astronaut, he flew one Gemini flight in earth orbit and commanded the *Apollo 15* trip to the moon). Scott was making a simulated X-15 landing approach for training and evaluation purposes when the aircraft suddenly lost partial power and began to drop rapidly. Both pilots realized that the F-104 would hit extremely hard. They made opposite decisions that saved both their lives.

Dave Scott elected to stay with the aircraft while Mike Adams ejected. Adams pulled the ejection D-ring just at the moment the F-104 slammed into the runway, collapsing the landing gear. If he had punched out before impact, the rapid rate of aircraft descent would have counteracted the upward thrust of the ejection seat, resulting in insufficient time for his parachute to blossom and safely carry him to the ground. If he had delayed ejecting even a fraction of a second after runway contact, he would have been crushed when the aircraft's General Electric J79 jet engine slammed forward into his rear cockpit. Adams' parachute successfully deployed seconds before he hit the ground.

Due to the extreme forces encountered on runway contact, the aircraft broke apart just between the front and rear cockpits. Since Scott stayed with the aircraft his cockpit tumbled out into the desert as the rest of the F-104 disintegrated and burned. Scott's ejection seat had partially sequenced automatically due to runway impact, locking his feet in the stirrups. If he had attempted to eject, Scott would have been killed. He simply opened the canopy, stepped out of the cockpit uninjured and saw Adams waving at him as he descended in his parachute.

The cool professionalism Adams displayed in this accident impressed the test pilot school instructors, and he went on to become the top student in his class. Adams had waited for the precise time to act, a technique that unfortunately did not serve him later.

To keep proficiency in flying at night and in instrument conditions, pilots scheduled T-33s for cross-country flights. Our test aircraft were never flown on proficiency flights—they were strictly used for test missions.

Soon after I had arrived at the Fighter Branch, Mike Adams scheduled a T-33 for a weekend trip to visit his brother George in Sacramento. Adams invited me to go along with him. Brother George owned a pizza parlor, so we expected free pizza and all the beer we could drink.

We had an uneventful flight to Sacramento and enjoyed pizza and beer with Mike's brother. On the return trip back to Edwards, we began to compare stories about cross-country flights we had each taken that had gone awry with aircraft malfunctions. I relayed to him my F-101B Voodoo story about my landing gear and armament problems, which resulted in my standing at attention explaining to my squadron commander why I did not return home with my two live missiles.

Not to be topped, Adams had a story of his own. He had taken many cross-country flights to Sacramento to visit brother George. Not all the trips were for free pizza and beer—his mother also lived in Sacramento.

On one of the trips he flew solo with no one in the rear cockpit. During his visit he found a perfect bird dog he planned to use for hunting. The problem he encountered was that he didn't have transportation for the dog back to Edwards. Then he remembered the empty backseat in his T-33. With just the right arrangement of seat belt and shoulder harness the dog became his copilot on the way home, making that dyslexics' bumper sticker "*Dog is my copilot*" a reality in this case.

Shortly after leveling off at 30,000 feet on his return flight, the cabin pressurization failed on the aircraft. Adams was wearing a standard Air Force oxygen mask attached to his helmet and activated his oxygen system. He started to descend slowly to about 20,000 feet. It was now fully dark and he had to cross the Sierra Nevada range with peaks not visible. As best as he could tell in twisting around in his seat to observe his copilot, the dog was doing fine.

After landing, he unstrapped the dog and removed any evidence that a dog had been in the seat. Dogs were certainly not authorized passengers in military aircraft. Fortunately, the transit alert crew was busy recovering another aircraft and the dog was removed without notice.

Unfortunately, due to the prolonged flight with low cabin pressure, the dog had suffered from the lack of oxygen and had become hypoxic. Adams managed to bring the dog back to consciousness, but it had difficulty staying on its feet. In the weeks to follow the dog made a partial recovery, but it lost the complete use of one leg. It hobbled around on just three legs, never becoming the hunting dog he desired. Mike Adams named the dog Tripod.

Chapter 10

Phantom Foibles

The story of the F-4 Phantom II began in the early 1950s when McDonnell Aircraft Corporation submitted an unsolicited proposal to the Navy for a successor to the Phantom I, the first jet-powered carrier fighter. In October 1954 the Navy awarded McDonnell a contract to build two prototypes of the AH-1, a twin-engine, two-seat, all-weather attack aircraft. Seven months later, the Navy decided the new plane should be a missile-armed fighter. They redesignated it the F4H-1 and placed an initial production order. The first one flew in 1958.

After building ten F4H-1s (later redesignated the F-4B), the Navy modified the eleventh production item for a ground attack role as well. That plane was called the F4H-1F, which in 1962 became the F-4A. The latter design proved so useful to both the Navy and Marine Corps that existing F4H-1s (F-4Bs) were later retrofitted to that standard.

The Navy lost no time in setting a raft of records with the F-4B, including world speed records of 1,606.3 mph at Edwards and 902.72 mph over the hazardous 3-kilometer low-level course, a Los Angeles–New York record of 2 hours 48 minutes, a speed record over 100 km and 500 km closed courses, a sustained altitude of 66,443 feet, a zoom altitude of 98,557 feet, and eight time-to-climb marks between 3,000 and 8,000 meters.

Gen. G. P. Disosway, commander of the Air Force's Tactical Air Command, decided to buy the Navy plane. "It did not take long for us to decide on the Phantom," he recalled. "For one thing, it had most of the speed and altitude records. While that may have seemed colorful to some, to the tactical fighter pilot performance equals survival. He needs an aircraft capable of meeting the enemy on his own terms, at the enemy's speed or better, and at his altitude."

Originally the Air Force version was called the F-110A. When the Defense Department set up a joint aircraft designation in 1962, the Air Force Phantom became the F-4C. The plane's beaklike nose drooped out of a barrel belly. Its wings bent up at the tips. Its tail fin was about as graceful as a homemade weathervane. A little kid seeing the plane for the first time screamed excitedly,

"Look, Mom, they got the tail on upside down!" Even Air Force pilots called the plane the Double Ugly because of its massive size and unusual shape.

While McDonnell was tooling up to produce the F-4C, the Air Force borrowed twenty-seven F-4Bs from the Navy to begin training TAC crews. These planes were assigned to the 4453rd Combat Crew Training Squadron at MacDill Air Force Base, Florida. One of the first pilots to be checked out there was Col. Frank K. Everest Jr., also known as the Fastest Man Alive because of his speed record in the Bell X-2 at Edwards and a book he published by the same name. Everest soon became commander of the 4453rd, later moving with it to Davis-Monthan Air Force Base, Arizona. As the war in Vietnam escalated, more combat crew training squadrons were formed. Soon eight F-4s were lost in TAC training squadrons due to stall/spin loss-of-control accidents. The Flight Test Center at Edwards was tasked to determine the cause and to recommend a solution.

F-4C serial number 63-7409, the third Air Force aircraft built by McDonnell, had been used at Edwards in the initial stability and control testing. The aircraft was fully instrumented and given to me for flight testing. Flight test engineer Leroy Keith and I wrote a test plan and prepared to start testing. For the initial flight I had the Phantom loaded at a forward CG. At 10,000 feet and 300 knots I activated the instrumentation system, pulsed the stick fore and aft, and recorded the aircraft response. I slowed the speed on the F-4C and evaluated its near-stall and full-stall characteristics in straight and level flight. This was followed by a series of wind-up turns at speeds from 200 to 550 knots. Wind-up turns were started in stabilized level flight, followed by rolling into a bank and allowing the nose to drop, gradually increasing the G force without changing the thrust setting, and descending to hold a constant airspeed. I maneuvered to heavy buffet, which occurred at approximately 18 units on the angle-of-attack gauge. Due to yaw oscillation, wing rock became so bad that precision control of the aircraft was impossible above 20 units. Above 22 to 24 units, any aileron input produced a rapid roll off in the direction opposite that commanded by the lateral control. McDonnell engineers designed a rudder shaker to be activated at 22 units, but it was completely masked by heavy airframe buffet. The only flight control left for maneuverability at high angle of attack was the rudder. In the first month I flew six test missions.

On following flights I performed the same tests with the F-4C in a mid-CG and recorded similar results. The aircraft could be maneuvered, but the pilot had to carefully monitor his control inputs and be alert to counteract stray aircraft motions.

Flying near the aft CG was extremely dangerous. Not only was the stall masked by wing rock and heavy buffet, but the stick forces were so light it was easy to overshoot the desired G. In some wind-up turns, after I had started to apply Gs, I actually had to push forward on the stick to keep the G force from increasing and the F-4C from pitching up. As a result of this terrible characteristic, I came to the conclusion that some inexperienced student pilots had maneuvered the Phantom through airframe buffet and into a full stall, ending in their loss of control and a crash.

After the test flights were completed I came up with a method to move the CG forward. If the external tank fuel was initially restricted from being transferred into the Phantom's two aft internal fuel cells, the plane was in better balance. Test engineers and I recommended that several cockpit displays be modified to alert the pilot when he was nearing high angle of attack. We also wrote several cautions for the pilot's flight manual. Over the next five months I flew about thirty-five test missions in the F-4C. I didn't think much of the aircraft.

In the first two years of combat in Vietnam, the casualties among the first F-4C squadrons reached almost 40 percent of the aircraft deployed, for a total of fifty-four aircraft. Some were lost to ground fire, but many were lost in stall/spin accidents at low altitude. During close-in dogfights, when pulling high Gs or when flying at high angles of attack, it was very easy for the pilot to lose control of the F-4C, especially if it was carrying external stores. Recovery from a spin at an altitude below 10,000 feet was essentially impossible, and the only option left for survival was for the crew to eject.

Following my tour at Edwards, I flew 188 combat missions in Vietnam in the Douglas A-1 Skyraider. With the call sign Sandy, the A-1 was used to rescue downed Air Force and Navy aircrew in North Vietnam and Laos. I also flew as a forward air controller (FAC) in the A-1 and marked ground targets with white phosphorus rockets for F-4 and F-105 jets. It was my experience that pilots flying the F-4 had great difficulty precisely hitting ground targets. It was unusual for them to get a direct hit. I personally attribute this poor bombing performance to a marginally stable aircraft being flown at an aft CG by a low-time pilot.

Many F-4 aircraft crashed during these air strikes. We Skyraider pilots always chalked the losses up to enemy ground fire. During the course of the Vietnam War, out of a total loss of 2,254 Air Force aircraft, about 20 percent of them were Phantoms. From my flight test experience it is now my belief that more F-4s were lost due to out-of-control maneuvering by the pilot than were downed by the enemy. F-4 losses would have been fewer if the Phantom combat pilots had been briefed by someone who said, "Don't kill yourself!" like the general at Edwards did to me.

When I first set foot in Test Operations in 1965 I also recognized the smiling face of Capt. Chuck Rosburg. We had first met about two years earlier at Brooks during the five-day physical examination, which we both took as a requirement for entrance into the test pilot school. Rosburg was the broad-shouldered pilot who had reminded me of the Alley Oop character then in newspaper cartoons. I had outlasted him a few minutes on the Brooks treadmill test, but running was not his specialty. Rosburg's sport was handball.

Chuck Rosburg could not be beaten at handball. He became the base champion at Edwards and even competed in military tournaments around the country. Not only could he beat anyone in singles, sometimes we scheduled two people (as in doubles, two against one) to play against him. Rosburg would still win—he was in a league of his own. He created a crowd at the gym, as many people took time off from work just to watch him play. Rosburg was the best pilot-athlete I knew in all my years of flying.

Rosburg, four years older than I, was a bomber pilot by military education and training. But he was a fighter pilot when he was at parties, which he loved to attend. Short and feisty, he tended to lead with his chin. People were immediately attracted to Rosburg at parties because he was entertaining, full of fun, quick-witted, and the center of attention when he told stories. He liked to tease everyone and had a very assertive attitude, one that said, "Here we go—it's party time!" Maybe that explains why he and his wife Shirley had five children.

At Test Operations, Rosburg was a member of the Special Projects Branch. Pilots in this organization flew the Lockheed U-2 Dragon Lady on test programs. The U-2 project was initiated in the early 1950s by the CIA, which desperately wanted accurate information about the Soviet Union. Overflights of the Soviet Union with modified Air Force bombers started around 1951, but they were vulnerable to antiaircraft fire and fighters, and a number of them were shot down. It was thought a high-altitude aircraft such as the U-2 would be hard to detect and impossible to shoot down. Lockheed Aircraft was given the assignment with an unlimited budget and a short time frame. Its Skunk Works, headed by Kelly Johnson, performed remarkably, and Tony LeVier flew the first flight in August 1955. Kodak also developed new aerial cameras, which worked well. The U-2 made its first overflight of the Soviet Union in June 1956.

A Pratt & Whitney J57-P-37A turbojet engine powered the U-2. With 11,200 pounds of thrust, this power took the 80-foot wingspan aircraft to over 75,000 feet. At a maximum takeoff weight of 16,000 pounds, including about 4,000 pounds of fuel, the U-2 cruised at 500 mph for a range of 2,200 miles.

The aircraft first came to public attention during the U-2 crisis when pilot Francis Gary Powers was shot down over Soviet territory on May 1, 1960. Powers took off from Incirlik Air Base, Turkey, did a diversionary maneuver, and then continued the mission as planned until he reached Sverdlovsk, Russia. While he was on the photo run at 67,000 feet, the Soviets launched a volley of fourteen SA-2 surface-to-air missiles (SAMs) at Powers' aircraft. Although the SA-2s could not achieve the same altitude as the U-2, the aircraft disintegrated in the shock waves caused by the exploding missiles. Soviet authorities subsequently arrested Powers after he successfully ejected from the plane and held him on espionage charges for nearly two years.

On October 14, 1962, it was a U-2 from the 4080th Strategic Reconnaissance Wing that photographed the Soviet military installing nuclear warhead missiles in Cuba, precipitating the Cuban Missile Crisis. However, later in the crisis, another U-2 was shot down over Cuba, killing the pilot, Maj. Rudolph Anderson. The Soviet development of SAMs that could reach the U-2, the type that shot down Powers and Anderson, prompted the development of a very fast, very high flying reconnaissance plane, the CIA's A-12 and then the Air Force's SR-71.

Chuck Rosburg got to fly some of the last flights of a program titled Missile Detection and Alarm System (MIDAS). An electronic device installed on board the U-2 was used to provide early warning of a Soviet nuclear attack by detecting and tracking rocket plumes from an enemy missile's boost phase as it ascended into the stratosphere. It was an ambitious project, given the state of the art in infrared technology at that time and the huge distances over which the sensors would be expected to detect missile activity. Rosburg spent many flight hours at high altitude suited in the MC-3 partial-pressure suit, the same type of suit I wore in the F-101B Voodoo and Francis Gary Powers wore over Russia.

One U-2D was modified with a second seat for a navigator to operate the electronic system. They chased the fiery rocket plumes from U.S. missile launches at Vandenberg Air Force Base and Cape Canaveral. The U-2D used for the tests was nicknamed "Smoky Joe." This name was an acknowledgment of their role in monitoring rocket plumes. Smoky Joe was a rotund, cigar-smoking Indian character from the *Brave Eagle* TV show, and the Special Projects Branch adopted him as their mascot. A large painting of Smoky Joe greeted visitors to the unit's office, which was tucked away in a corner of the flight operations building, and a smaller version appeared on the tail of their U-2.

Another project that used the U-2 for testing was titled High Altitude Clear Air Turbulence (HiCAT). HiCAT studied the occurrence and intensity of high-altitude turbulence so that future fast and high-flying aircraft like the supersonic transport (SST) aircraft could be designed to safely fly through it. Lockheed designed and built instrumentation and a special gust probe with fixed vanes that were stuck on the nose of the U-2. It was painted red with white stripes, and so it inevitably became known as the Barber Pole. The U-2 was deliberately flown into regions where clear air turbulence (CAT) was predicted, resulting either from jet streams, weather fronts, temperature inversions, or mountain waves. Years earlier, the first CIA U-2 pilots would have been aghast at the idea of looking for CAT. Their impression was that the U-2 was a very fragile aircraft, not built for turbulence.

Search flights looking for CAT were made all over the North American continent, flying out of Edwards, Hickam in Hawaii, Ramey in Puerto Rico, Hanscom in Massachusetts, and Elmendorf in Alaska. Rosburg made many test flights in the U-2 on both programs.

During long excursions away from Edwards, the U-2s were often accompanied by a converted JB-47E Stratofortress bomber, the same type of aircraft Rosburg had flown on nuclear alert in Strategic Air Command for six years before he attended test pilot school. The B-47 was used as an escort to aid in navigation and communication relay, in the same way that SAC and CIA U-2s were trailed by Boeing C-135s, especially on long over-water flights.

Capt. Jerry Tonini, one of my classmates from test pilot school, had also flown the B-47. Like Rosburg, Tonini had spent many years in SAC on nuclear alert in the B-47 but now was assigned to the Bomber/Transport Branch of Test Operations flying the B-47, C-130 Hercules, and the new C-141 Starlifter.

One day Chuck Rosburg and Jerry Tonini were scheduled for a proficiency flight in the B-47. This was not a test flight; the two of them planned to make several instrument approaches and end up with each pilot flying several touch-and-gos. I was invited to go along. Since I was not qualified in the B-47 and did not know any of the normal or emergency procedures for it, I assumed I would sit in the bombardier seat in the nose of the aircraft. The B-47 carried a three-man crew, with pilot and copilot sitting in tandem, unlike civil and military transports, where pilots sit side by side.

Rosburg and Tonini decided to let me fly from the front seat. While Tonini started the six jet engines from a crawlway connecting the front and rear cockpits, I strapped in with a shoulder harness and lap belt. I noticed how much higher above the ramp I felt in the B-47 than in the jet fighters I was flying at that time.

As I lined up on the runway I pushed the throttles full forward on all six engines to obtain maximum power and made the takeoff. After we were airborne, I realized I had forgotten to remove the safety pins that armed my ejection seat. Rosburg and Tonini had forgotten theirs also. Unlike the strict rules and procedures followed in SAC, Rosburg and Tonini did not use a checklist. We played the flight by ear.

After accelerating to a high speed of about 500 knots at low altitude, Rosburg showed me how the aerodynamic forces lifted the wingtips many feet above their height when on the ground. Then he turned the aileron hydraulic boost off. The flight controls were now reversed. If I attempted to apply aileron motion in one direction, the slender wing would warp, resulting in an opposite effect. When I moved the yoke to the right, the B-47 banked left. When I moved the yoke to the left, the B-47 banked right. This was a very dangerous condition if you were flying in weather or at night and the aileron boost failed. I felt uncomfortable flying an aircraft in which I did not know the flight limitations. Red warning lights were glowing on the instrument panel and heavy forces were required to fly with controls reversed.

Landing a six-engine heavy bomber with bicycle-type landing gear was also a challenge. The touchdown speed and aircraft pitch attitude were very critical. If a pilot missed on either count, the aircraft porpoised and could crash. My landings were good under the circumstances, but I was glad to make my last landing and taxi the B-47 back to the Edwards ramp.

I realized that bomber pilots have their hands full flying a huge and complex aircraft. Rosburg and Tonini must have had a lot of confidence and proficiency in the B-47 to let someone as unqualified as I was fly such extreme flight conditions from the front seat. I wasn't sure I would let an unqualified pilot fly that close to the limits in a fighter.

Later that year, I flew to McClellan Air Force Base, just north of Sacramento, to pick up and return an F-5 that had just completed a major depot-level maintenance inspection. Rosburg flew a T-33 to McClellan and I rode along in his backseat as a passenger.

As I preflighted the F-5, Rosburg followed me around the aircraft. He said his goal was to complete his tour flying bombers and the U-2 and eventually start flying fighters. He was thrilled to sit in the F-5 cockpit on the ground and evaluate its superb visibility and cockpit design. I thought the F-5 cockpit design was the best that I had ever seen in a fighter aircraft. As aggressive as Rosburg was on the handball court and at parties, I could easily picture him as a fighter pilot.

"Hold your position," crackled the Edwards Command Post over the UHF radio. "We are changing your mission. The number two XB-70 Valkyrie has crashed."

The date was June 8, 1966, and I was acting as an instructor pilot in a T-38 Talon. My student, sitting in the front cockpit, was a SAC colonel obtaining some supersonic flight experience in the T-38 before checking out in the SR-71. We had preflighted the aircraft and were preparing to start engines when we received that chilling call.

It did not surprise me when I heard that an XB-70 had crashed. In the year and a half that I had been acting as a safety chase pilot flying a T-38, F-4, or F-104, the XB-70 had had many close calls. While the official Air Force name for the XB-70 was Valkyrie, North American Aviation, builder of the aircraft, called it the "Great White One." Because of all the mechanical problems the aircraft experienced, we safety chase pilots called it "Cecil, the Seasick Sea Serpent." It did resemble a sea creature with its long, thin neck and curved body, and it became sick on every test flight that I can recall.

Chase pilots routinely accompanied the XB-70 on every flight. We flew a loose formation position about one hundred feet abeam the test aircraft and looked for any unusual or dangerous conditions and called them to the attention of the XB-70 test pilot. We had to be alert—aircraft problems occurred on every flight, some very serious.

"One of these days Cecil will stub its toe and go down," I had previously predicted to other pilots.

Little did I realize then the strange and ironic sequence of events that would occur to make that day unforgettable—a day I feel was the darkest day in Edwards flight test history.

Moments after being told my mission would be changed, the Command Post told me that the XB-70 had crashed about thirty-five miles from Edwards, just north of Barstow, California, and directed us to take off and orbit over the crash site acting as a radio relay. We would relieve a T-33 that had been above the scene giving information on the unfolding events. It was running low on fuel and needed to return to Edwards. Additionally, a helicopter had been scrambled from Edwards to recover survivors, but it would not arrive for another ten or fifteen minutes. We should orbit overhead and relay information since the Command Post could not maintain line-of-sight radio contact with the helicopter while it was on the ground.

As soon as we were airborne in the T-38, I spotted black smoke from the XB-70 crash site drifting up to around 10,000 feet in the clear, blue Mojave sky. As the rescue helicopter approached the crash scene, we relayed a report from the Command Post to the chopper pilots to look for evidence of another downed aircraft, a NASA F-104N. Parts of the XB-70 were still intensely burning, a picture that is still fresh in my memory even after forty-one years. I personally knew all the XB-70 pilots and many of the NASA pilots. Death was becoming a common visitor in the test community because of so many aircraft accidents; I feared its presence again.

Orbiting over the smoking ruins, I recalled the first time I saw the XB-70. It was in 1964 when I was a student in the test pilot school. Visiting the North American Aviation facility at Palmdale, California, I watched the incredible XB-70 roll out into the aviation spotlight of the world. The long-secret plane looked like a hooded cobra inside the shadows of the huge hangar where final assembly had been completed. The majestic plane, covered with a special white paint, was designed to ride on its own shock wave. Its needle-like nose was shaped to push aside air beneath the triangular wing. The tips folded down 65 degrees to provide the same type of lift as the wave under a high-speed boat.

The aircraft was gigantic, 196 feet long with a 105-foot wingspan and twin 30-foot vertical tails. Even the air ducts, which supplied air for six General Electric J93 engines, were huge, measuring 7 feet high and 80 feet long. The engines themselves were in the 30,000-pound class, gulping air through twin ducts controlled by a very complex and troublesome intake system. The eight landing-gear tires appeared to be sprayed with an aluminum-colored paint, giving the Valkyrie a science-fiction image. Weighing 550,000 pounds, the XB-70 was in a class by itself.

Now my mind flashed back to the many flights I spent chasing the Valkyrie as it performed flight test maneuvers. Due to hydraulic, electrical, or other mechanical failures occurring on practically every flight, several safety chase aircraft flew with the XB-70 during its test mission. A T-38 usually accompanied the XB-70 on takeoff and landing, confirming that the landing gear worked properly. Due to the complexity of the XB-70 landing gear operation, the takeoff safety chase crew performed an in-flight pickup. The takeoff chase pilot launched first, climbed over Rogers Dry Lake, and pulled up into a downwind pattern opposite the heading of the XB-70, which was awaiting start of takeoff roll on Runway 04.

When the XB-70 pilot called out "Rolling in 30 seconds," the T-38 pilot timed the start of his 180-degree descending turn so as to be directly abeam the XB-70 as it lifted airborne. Standard practice called for the XB-70 test pilot to delay gear retraction until the T-38 pilot moved in under the aircraft and was in a position to watch the complex gear retraction motions. One of the other XB-70 test pilots frequently rode in the rear cockpit of my T-38 and visually confirmed correct gear retraction.

Fourteen months before the accident, the eighth flight of the first aircraft saw the last of the four pilots, North American's Van H. Shepard, fly the XB-70. Shepard, forty-two, was from Monroe, Louisiana. Occasionally he flew in the backseat of my T-38 like the other XB-70 pilots. An important milestone was reached on this eighth flight of the number one XB-70 on

March 24, 1965, when Al White and Van Shepard took the XB-70 on its first double-sonic flight. They went to 56,000 feet on that sortie, quickly reached Mach 2.14 (1,365 mph), and flew beyond Mach 1 for seventy-four minutes, forty minutes of which were in excess of Mach 2. That was the day Las Vegas, Nevada, learned about the XB-70. Because of a momentary glitch in the single TACAN (Tactical Air Navigation) system, the flight led to an inadvertent "booming" of Sin City in the early morning. The supersonic bomber and its B-58 chase plane passed twice over the city, once at Mach 1.9 and again at Mach 2.1, slamming down a series of heavy sonic booms that led to flooded telephone switchboards and set off a memorable civic uproar.

Just three months before the crash of the second XB-70, Van Shepard and Col. Joseph F. "Joe" Cotton, USAF, had a close call as they flew the number one aircraft on its thirty-seventh flight. Cotton, forty-four, from Newcastle, Indiana, learned to fly in World War II. He flew the B-29 and B-36 and then attended the Empire Test Pilot School in England. There he flew the Sea Fury, Meteor, and Vampire jets. After graduation he tested the Boeing B-47 and B-52 and the Convair B-58. He was now the commander of the XB-70 test force.

Performance tests were on the flight plan for Shepard and Cotton. But soon it was a different performance being tested. Halfway through the planned flight, both hydraulic systems, primary and secondary, failed. Shepard quickly brought the Valkyrie home, and Cotton extended the gear for landing. The green landing gear indicator lights did not come on; this was followed by a call from the chase plane that there was trouble with both sets of main gear. On the left side, the gear hadn't fully lowered before rotating to meet the direction of travel, leaving the rear wheels higher, rather than lower, than the front set of wheels. The right side gear was in worse condition—it hadn't lowered at all before rotating. Even more alarming, it hadn't rotated completely in line with the direction of travel, although it was close. The emergency backup system failed to correct the problem, and with overall control of the aircraft rapidly coming into question, engineers on the ground had to think quickly, or face losing an aircraft that was a half-billion dollar investment.

After what must have seemed like hours to the pilots in the cockpit, engineers on the ground called with their plan. Shepard should land the Valkyrie on Rogers Dry Lake, so there would be plenty of room to ease it to a stop. The engineers thought that, on touchdown, the left side gear would level itself out, the sheer weight of the XB-70 forcing the gear into its normal position. As for the right gear, being behind the centerline of the main strut, it was unlikely that the gear would level out, but hopefully the landing would at least cause the gear to finish swinging into the direction of travel, and the wingtip would still clear the ground, although the right side would be much lower than the left.

Shepard gingerly set the XB-70 down on the lake bed, and each main gear did what the engineers expected. The plane was on the ground and rolling, but it wanted to turn sharply to the right—threatening to ground loop. A ground loop, although not likely to be fatal to the pilots, would nevertheless destroy the huge long-nosed aircraft. Shepard kept applying power to the number six engine (the farthest engine on the right side) to help keep the XB-70 somewhat straight. But at the same time he couldn't use too much power, or the plane would never stop!

Finally, after rolling almost three miles, the XB-70 completed its landing run. This run, when viewed from above, looked like an upside-down letter J because the XB-70 had not just swung over half a mile to the right side, but by the time it came to a stop the plane had turned 110 degrees, almost pointing in the direction it had come from. Only the huge size of the lake bed made this landing possible—anywhere else the XB-70 would have been far off the runway and likely would have run into buildings and hangars. Rogers Dry Lake had saved another rare experimental aircraft.

While flying safety chase, I observed landing gear malfunctions many times. I saw the huge tires on one gear arranged perpendicular to the ground—an eerie sight. Once I even saw the entire landing gear rotated 90 degrees away from the fuselage centerline, a condition that would require the test pilots to eject if it could not be corrected. There was always a sense of impending doom as one flew near this beautiful Mach 3 bird—disaster could be only moments away.

Another time, while I was flying in the backseat of an F-104D piloted by fellow fighter test pilot Capt. Joe Stroface, the XB-70 pilots asked us to fly directly under the XB-70 and visually check a landing gear linkage in the wheel well. The F-104 had a T-tail extending about eight feet higher than the pilot's head, so vertical clearance of our tail and the underside of the XB-70's lower wing would be a very critical factor. Stroface deftly flew the F-104 not more than twelve to fifteen feet directly underneath the XB-70. Meanwhile, I was absolutely terrified without my own hands on the flight controls and throttle to assure myself that we would not collide. I pressed my body as low in the corner of the upward ejection seat as I could—knowing full well that this position could not possibly save me if we did collide. But taking any action at that moment made me feel better.

Early in the XB-70 flight test program, the J93 engines could not be restarted in flight unless the aircraft was flown to at least 0.8 Mach, an excessive speed to have as a standard procedure. North American Aviation and General Electric engineers suspected that an irregular fuel pattern was to blame. The engineers requested that I fly a T-38 with a North American Aviation photographer in the rear cockpit and attempt to fly the aircraft directly behind the engine exhaust nozzles so that photographs could be taken of the fuel pattern. I flew the nose of the T-38 to within fifteen feet of the engine exhaust nozzles and slightly low, so the photographer could aim his camera directly above my flight helmet into the nozzles. Flight procedures had not been published nor briefed to safety chase pilots on special precautions to be taken in flying very close to the XB-70. We simply relied on close verbal communications with the XB-70 test pilots and used whatever flight control movements were required to slide into position and stay there. We were considered part of the test pilot fraternity and expected to use good judgment. We got the photos.

To complete flight tests at Mach 3, the XB-70 flew a huge circular flight path covering California, Nevada, Idaho, Utah, and Arizona. Frequently a Convair B-58 Hustler, with another XB-70 test pilot at the controls, would fly a Mach 2 circle over the ground staying inside the XB-70's path. That way another test pilot with intimate knowledge of the XB-70 could quickly join up and render assistance if needed. Also, an F-104 was pre-positioned at Mountain Home Air Force Base, Idaho, or Hill Air Force Base, Utah, ready to join up in formation if the XB-70 needed to land at either base or one of the dry lake beds in that area. A massive armada of support aircraft was required for each XB-70 flight, similar to the X-15 rocketship missions.

The most demanding and dangerous personal experiences I encountered with the XB-70 before the accident of the second aircraft involved flying formation in supersonic flight. North American Aviation management wanted photographs of the XB-70 with the wingtips at their full down position of 65 degrees. The aircraft took off with the tips at 0 degrees, or level. As the plane accelerated to 500 knots the tips were lowered to 30 degrees. Above Mach 1.2 the tips were placed full down. Aircraft flying at supersonic speeds create shock waves, which can be heard by an observer on the ground. The pilot in the supersonic aircraft does not hear this shock wave. He just feels slight changes in his flight controls and sees a momentary hesitation in his airspeed indicator around Mach .95 until it jumps to over Mach 1. The North American photo would have to be acquired near the shock wave.

The XB-70's shock wave had a great impact on a chase plane flying in formation both from a visual and aerodynamic perspective. Portions of the shock wave were visually observed as an optical distortion of the XB-70 airframe. In high school physics, we see the effect of placing a pencil in a glass of water. Due to the difference of the index of refraction between air and water, the pencil appears to be bent or displaced. This same effect occurred with the Valkyrie—the plane was not exactly where I visually observed it.

A shock wave came off both the XB-70's nose and tail, gradually bending back at a greater angle as the plane accelerated faster and faster. It was important for the pilot in the photo chase plane to fly forward of the shock wave created by the XB-70's nose, thereby flying in undisturbed air. If he got behind the shock wave, the F-104D, with tip tanks installed in the tandem cockpit version, did not have enough engine power to punch back through the wave. Because this shock wave is actually a cone surrounding the supersonic XB-70, a chase aircraft will also experience a different effect on the wing nearest the XB-70 while flying in the shock wave. This disturbance would cause the chase plane's nose to oscillate, becoming very "squirrelly." Many chase flights were flown before we got that perfect supersonic photo.

As we circled over the crash scene in the T-38, I watched the helicopter land and recover North American Aviation's chief XB-70 test pilot, Alvin S. "Al" White. White, forty-eight, was born in Berkeley, California, and had joined the Army Air Corps in 1941. He first served as an advanced flying instructor before going overseas. White spent two years in the European theater flying the P-51 Mustang in six major campaigns, including the Battle of the Bulge. He earned an engineering degree from the University of California in 1947. White then joined North American Aviation as an engineering test pilot and participated in the flight testing of the F-86 Sabrejet, the F-100 Super Sabre, the F-107, and the Sabreliner business jet. White was severely injured in the accident with back and arm injuries, but he survived. However, he never flew the XB-70 again.

The helicopter crew reported they had found one fatality, and we relayed that message to the Edwards Command Post. Later I learned that the pilot killed was Maj. Carl C. Cross, forty-one, from Chattanooga, Tennessee. Cross was an Air Force bomber/cargo test pilot flying his first XB-70 flight as copilot. He earned his bachelor of science degree through the Air Force Institute of Technology program at Wright-Patterson Air Force Base, Ohio, and attended the test pilot school in 1960. Cross was a quiet, unassuming type, just back from a combat tour in Vietnam. Ironically, after a year dodging ground fire in the war, he was killed in the skies over the Mojave Desert. His death was a tremendous loss to those of us who knew him at the Flight Test Center.

Several months before the accident, I had flown as copilot on a Lockheed C-141 Starlifter helping Cross perform "speed-power" performance tests. He had allowed me to make several touch-and-go landings in the Starlifter at the end of the mission. This was a tremendous thrill for me as a fighter pilot, a chance to fly what was then the most modern and advanced Air Force military transport. Follow-on versions of the C-141 flew for over forty years but have now all been retired.

Only two XB-70s, serial numbers 62-0001 and 62-0207, were ever built. The second and more advanced model was the one involved in the accident. As a result of this mishap, the remaining XB-70 was grounded pending results of one of the most extensive accident investigations ever performed at the Flight Test Center.

The final accident report stated that the pilots of the XB-70 had completed their scheduled test flight. General Electric (maker of the Valkyrie jet engines) had contracted for Clay Lacy to take publicity photos from his civilian Learjet at the end of the flight. The photos would show four fighter-type aircraft (the F-5, F-4, T-38, and F-104N), all equipped with General Electric engines, joined with the XB-70 in formation flight.

It was during this formation flight that the accident occurred. The F-104N, serial number N813NA, flown by Joe Walker, NASA's chief civilian test pilot at Edwards, who was on the XB-70's right wing, collided with the larger aircraft. The collision caused the F-104N to disintegrate and ripped off all the XB-70's left vertical tail and half of its right. Without the vertical tail to provide directional stability, the XB-70 flew quite a distance before it fell into an uncontrollable spin. From the wreckage of both planes, it was quickly determined what caused the accident. Why the accident occurred was never precisely determined.

When two aircraft fly in close wing formation, they disrupt the airflow on the wings closest to each other. Their wingtip vortices are slightly modified, causing each wing to produce a small increase in lift. With two similar aircraft in formation, each pilot unconsciously holds a little "aileron into" the other aircraft. When the two aircraft are of greatly different wing size and sweep angle, as the case with the XB-70 and F-104N, the relatively small F-104N experienced a much, much larger effect.

As automobile drivers, we sense the airflow off a large truck as we go around it on the highway at high speed. Our steering wheel must be slightly turned into the truck to maintain a straight course down the highway. As we pull in front of the truck we must remove this small steering angle to remain aligned with the road. We make these corrections automatically, sometimes without even being aware of it. Pilots don't look at their control stick any more than drivers look at their steering wheel. To maintain a safe position requires continuous visual contact with the other vehicle and smooth but quick adjustment of the steering wheel, or, in the case of an aircraft, the control stick.

Joseph A. "Joe" Walker, forty-five, from Washington, Pennsylvania, was a very experienced pilot, a NASA test pilot for twenty-one years. Walker earned a degree in physics from Washington and Jefferson College before entering the Army Air Forces where he flew the P-38 Lightning in World War II. Walker had flown almost every experimental rocket aircraft from the early Bell X-1 to the North American X-15, flying the X-15 to a world speed record of 4,104 mph and a record-breaking altitude of 354,200 feet. He had thousands of hours in the air and had flown practically every jet aircraft produced by the United States. In fact, Walker was scheduled to be the next pilot to qualify in the XB-70.

Evidently Walker was distracted for a split second and was weaving back and forth too close to the XB-70's right wing, or the strong wingtip vortices flowing off the XB-70 overpowered his ability to control the small-winged F-104N and his jet was pulled into physical contact. In either case, the destruction of his aircraft was total and his death immediate.

The black cloud from "Cecil" continued to darken the Mojave Desert. A couple of days after the accident, my friend Capt. Joe Basquez, an Air Force helicopter test pilot, was ferrying members of the accident board back and forth between the crash site and the Edwards flight line in a Bell UH-1 Huey. Basquez, thirty-one, was nicknamed "Jolly Joe" because of his magnetic personality and positive attitude. He had just graduated from the U.S. Naval Test Pilot School in Maryland and was assigned to Edwards to test helicopters. After landing on the ramp, Basquez shut down his jet engine and started walking away. Suddenly a gust of wind caused one blade to arc down, striking him in the head. His severe injury required medical evacuation to a naval hospital in San Diego. Basquez never flew again. Likewise, the General Electric engine photo flight was never attempted again.

Flying safety chase on the XB-70 Valkyrie was a thrilling experience, with never a dull moment. It was a challenge to fly far enough away to maintain a safe position and still be close enough to be of value to the XB-70 test pilots. As chase pilots we always flew with a great sense of caution and concern, never forgetting we were swimming through a great sea of air with a creature called Cecil, whose habits were not all known and whose very size could consume us in a flash—as it did my friends on that tragic day on the high Mojave Desert.

In between test missions, test pilots were scheduled for proficiency flights in the T-38. Two pilots were assigned to fly for an hour or two and encouraged to practice instrument flight. Few actual instrument approaches were ever made at Edwards because the weather was generally clear.

Capt. Joseph H. "Jovial Joe" Engle, thirty-three, and I were scheduled for an instrument proficiency flight with Engle sitting in the front seat and me in the rear. Engle was three years older than I, born and raised in Dickinson County, Kansas, just south of my home state of Nebraska. He attended the University of Kansas, enrolled in Air Force ROTC as I did, and came on active duty after a year as a flight test engineer with Cessna Aircraft Company. Joe was tall, thin, and had the exuberance of two pilots. He loved to play Ping-Pong and attempted to beat the senior test pilots at acey-deucey, a version of backgammon popular in the Navy.

At the time of our T-38 flight Engle had flown sixteen flights in the X-15 at speeds up to Mach 5.71 and an altitude of 280,600 feet. Engle got his Air Force astronaut wings by flying over fifty miles on his fourteenth flight in the X-15 and was now anxious to get accepted into the NASA astronaut program in Houston. Engle had a longing to make a flight to the moon. Other test pilots at Edwards advised him not to apply for NASA. They reasoned that he already had astronaut wings, that he had the best test pilot job in the Air Force flying the X-15, and that he would be treated as a new guy in Houston. But Engle had his eye set on going to the moon.

While I was making a practice instrument approach sitting under a fabric hood, which prevented my view outside the cockpit, Engle shouted, "I've got it," and shook the control stick. This meant he was taking control of the aircraft. Engle had spotted a flight of four F-4 Phantoms that had just departed the Edwards gunnery range. They were headed south, returning to their home base at George Air Force Base about thirty miles southeast of Edwards. Engle had flown the F-100 Super Sabre with the 309th Tactical Fighter Squadron at George before coming to Edwards. He always looked forward to rat-racing any fighter aircraft he saw going to or from George.

The flight of four F-4s was a sitting duck. They were flying a fingertip formation at about 300 knots and 5,000 feet. Engle fell in behind the formation at about a mile and pushed up the Talon's power. We were in the F-4s' six-o'clock position, unseen by them and closing fast. Doing about 550 knots, Engle flew directly under the lead F-4, probably only twenty feet below him. As he passed under the lead F-4 he made a maximum G, wings-level pull-up directly in front of them. The Phantom leader's heart must have stopped. The sight of a T-38 going vertical in front of him, generating a great amount of jet wash and throwing his flight into disarray, was probably the last thing he ever expected.

I was also caught off guard by Engle's high-G maneuver. I thought he was just going to give them a high-speed buzz job. The extreme Gs caught me totally by surprise. My head and shoulders slumped forward against the harness and I grayed-out, losing my peripheral vision. I felt my head bobbing up and down and was confused for a few moments. It was a strange feeling—I had never experienced anything like it before, not even in the Air Force centrifuge at Brooks.

I regained my bearings several seconds later and was very dizzy. It took me some time to realize I was flying in an airplane, that I was in a T-38, and that Engle was in the front seat. Gradually I regained my senses and was happy we were on our way back to Edwards to land. Joe Engle had given me the ride and scare of my life! I wasn't angry at him—as a pilot I should have been prepared for anything. I'm responsible for my safety and survival.

It was a pure accident—I was sorry I hit Maj. Dave Tittle in the groin with a handball. Sometimes we got carried away playing handball at the Edwards gym. All the test pilots were Type A personalities—we all wanted to win on the court and in the sky. After I graduated from test pilot school, I continued to exercise at the gym. Some pilots played tennis, squash, or golf. Since learning the fundamentals of handball five years earlier at Hamilton Air Force Base, I still preferred handball.

Tittle had flown with me two and a half years earlier when I made my first flight in the T-38 and had taught me how to go vertical in the Talon. Later in the course, we flew together several more times in the T-38. Tittle was an excellent instructor and a friend of mine; I certainly didn't want to harm him.

On the handball court, Tittle and I played against each other in doubles. Being left-handed I normally defended the left side of the court and my partner the right. A ball was hit to my side and looked like it would die in the back corner. Tittle, who was in front of me on the left side of the court, thought it would be impossible for me to return the shot. He made the mistake of turning around to face the backcourt in preparation for another serve. I managed to retrieve the ball and fired a bullet, which hit Dave squarely in the groin. He doubled over and dropped to his knees. The ball was dead, requiring another serve.

Years earlier I had been hit in the right eye from the same misjudgment. My cheek became swollen, restricting my vision for several days. I was medically taken off flying status and had to undergo several eye tests before the doctor would clear me to fly again. Physical injury to a pilot is serious business. But accidents just happened on the court; it was simply part of the game.

I felt sorry that my shot had injured Tittle. But I couldn't do anything about it—getting hit by the handball was a frequent occurrence. We went on to complete the game; I didn't know how badly I had injured him. Like Tittle, most players shook the pain off and continued to play. Only time would tell if he had been seriously harmed.

The day after the handball incident, October 4, 1966, I talked to Tittle at Test Operations. He didn't speak about his injury, so I assumed he was OK—the injury probably wouldn't affect his flying performance. Earlier in the year he had been on loan from the staff of the test pilot school to a position in the VTOL branch and was now testing the second XV-5A Vertifan, serial number 62-4506. Ryan test pilot Lou Everett had been killed in the number one XV-5A the previous year. Tittle was pleased to be part of the "XV-5A's Fan Club," made up of test pilots who had flown the plane. He told me he was the third Air Force pilot to fly the XV-5A. Maj. Phil Neale, who had been killed in the French Balzac in 1965, was the first to fly. Maj. Robert "Bob" Baldwin, an Edwards helicopter test pilot, also flew the plane. By October Baldwin had departed Edwards for a combat tour in Southeast Asia, and I moved into his former house in the base housing area. Bob Baldwin was later killed in Vietnam. The XV-5A Fan Club was a small and exclusive group.

Tittle remarked, "Some of the test pilots claimed the Ryan XV-5A flew like a helicopter. Others compared it with a high-performance jet." He added with a laugh, "They're both right."

Little did I anticipate the tragic circumstances that would occur the next day to Tittle in the XV-5A.

The XV-5A Vertifan was designed and built for the Army's Transportation Research Command by Ryan Aeronautical Company, San Diego, in conjunction with General Electric, developers of the aircraft's lift-fan propulsion system. The Army believed that VTOL aircraft could make a major contribution to the mobility needed for limited warfare. The XV-5A would be used for surveillance of the battlefield and rescue of downed aircrew. Freed from dependence on airfields, the Vertifan blended the flexibility of the helicopter with the speed performance of a jet.

The Vertifan was designed to take off straight up, dart away at high subsonic speeds, and return to land in an area no bigger than a tennis court (about twice the size of a handball court). It was a small aircraft with a length of 44 feet and a wingspan of 30 feet. With an empty weight of 7,000 pounds and a maximum VTOL takeoff gross weight of 12,300 pounds, the XV-5A carried a useful load (fuel plus payload) of 4,730 pounds. At 35,000 feet it could fly 475 knots (Mach 0.72) with a ferry range just over one thousand nautical miles.

The aircraft's distinctive feature was a set of counterrotating, five-foot fans submerged in the wings, which were driven by jet exhaust to provide lift for vertical takeoff, helicopter-like hovering, and vertical landing. Fan inlet doors were installed above the wing and fan exit louvers below the wing. For vertical flight the rotating fans provided columns of relatively cool low-speed air for lift. These fans, which were simply multibladed propellers, operated at 2,250 rpm, about the same rpm as a light plane propeller. The low velocity and low temperature of the fans' exhaust gave the XV-5A good unprepared landing capability. A smaller fan in the nose provided pitch control.

During hover the fan inlet doors were open and the exit louvers (whose shape resembled that of the slats on a Venetian blind) were in the vertical or open position. For transition to horizontal flight the exit louvers directed fan exhaust aft, producing lift and forward thrust. When wing-supported flight speed was reached, exhaust was directed out the tailpipe. The wing-fan doors and louvers then closed and the aircraft flew as a conventional jet aircraft until landing, when the sequence was reversed.

The total thrust of the two General Electric J85 jet engines, which were mounted high on the fuselage, was 5,316 pounds. This thrust was augmented nearly 300 percent by the twin fans that operated during vertical takeoff and landing. Contrary to the belief that VTOL capability automatically meant high fuel consumption, the fuel required for takeoff of a lift-fan aircraft such as the XV-5A was identical to that for vertical, short field, or conventional takeoff conditions. This was true because the jet engine would be operating over the same time period.

The Ryan portion of the flight test program started in May 1964 and was completed by January 1965. Test results showed that the Vertifan could operate off sod, alfalfa fields, plowed dirt, and the desert floor.

All of the erosion test flights were performed by Ryan test pilot Val Schaeffer, whose favorite theme was: "The XV-5A is not a hovering machine—it's a jet aircraft that can also land and take off from unprepared sites." Earlier in the XV-5A flight test program, Schaeffer had flown the aircraft in conventional jet mode at a speed of 526 mph.

Test results indicated that the combination of high-speed conventional flight and the ability to hover and operate from unprepared sites would make a modified XV-5A a remarkable rescue aircraft. The Vertifan would be able to escort attack aircraft, loiter near the strike area, and, in the event of an emergency, descend rapidly to hover, recover a downed crewmen, and then evacuate the area at jet speeds. Experience in Vietnam showed that speed of recovery was essential, since completion of the rescue before the enemy could organize a concerted effort to prevent it vastly improved the chances of success.

The proposed rescue version of the XV-5A would have a passenger compartment behind the existing two seat side-by-side cockpit, a right side access door for the passenger compartment, and a recovery hoist that extended through the door during rescue operations. The hoist, operated by an observer seated on the right side of the cockpit, would be capable of lifting two people.

Flight test instrumentation was installed in the right seat of the second Vertifan, so all test missions were flown solo. To test the rescue capability of the XV-5A, a helicopter-type rescue winch was mounted on the aircraft behind the cockpit and set up to be operated by the test pilot. The Vertifan had flown backward at speeds up to 25 mph and sideways at speeds up to 35 mph, paced by a car. A 235-pound anthropomorphic dummy was raised on a fifty-foot cable to within four feet of the hovering XV-5A.

The Vertifan also hovered at varying altitudes while a live subject walked up to and under the aircraft. All tests indicated that the proposed new XV-5A would make an outstanding rescue vehicle. At the time, more than 350 test flights had been logged by fifteen Army, Navy, Air Force, and civilian test pilots during 160 hours in the flight test program, fully proving the lift-fan VTOL technology.

On October 5, 1966, Dave Tittle was scheduled to perform advanced tests with a lightweight rescue sling attached to the winch cable. The flight took place in the second XV-5A at the Edwards South Base runway.

Tittle was hovering at 50 feet and lowered the cable. The lightweight sling was then attached. A human subject placed the horse-collar sling around his arms and back; Tittle lifted him a few feet into the air and then lowered the subject back to the ramp. They repeated this maneuver several times.

The test subject released the horse collar, and Tittle started to reel the cable back into the hovering aircraft. The XV-5A was in a slight forward moving turn while raising the sling. According to Ryan observers, the lightweight sling appeared to swing into the top of the left-wing fan inlet, causing a reduction of lift. Because lift was lost from the left wing, the Vertifan started to descend nose level but in a slight left bank. At the very moment the left landing gear made contact with the ramp, Dave Tittle ejected. The landing gear crumpled, incurring structural damage, but the aircraft remained intact.

As Tittle started up the rails in his ejection seat, the left landing gear struck the ramp. With the left gear making contact with the ground a fraction of a second before the right, the XV-5A rolled to the right, then the right gear made contact. As Tittle's ejection seat reached the top of the rail, a sideward torque was placed on it. After leaving the aircraft, his seat tumbled sideways to the left several times, never gaining more than twenty feet altitude. This motion prevented Tittle from separating from the seat and the parachute from deploying. Dave Tittle died in the Edwards base hospital a short time later.

I was extremely sad to lose a close friend, someone who had been my instructor in the school and concerned with my safety.

This strange tumbling motion can be replicated by placing a pencil on a table with half of it extending over the edge. Then swing your hand down over the edge of the table, striking the portion of the pencil exposed. The pencil will tumble to the side.

Had Tittle ejected a fraction of a second earlier, his ejection seat thrust would have had to counteract the downward velocity of the Vertifan's sink rate. Still, he might have survived. On the other hand, had Tittle waited to eject until after ground impact, like Mike Adams did in the F-104, he would not have the downward velocity to counteract, but risked the XV-5A coming apart and damaging his ejection seat. Tittle had only a split second to make a critical decision. Only he knew what the situation looked like from the cockpit.

Unfortunately, VTOL aircraft have their flight controls directly influenced by engine power. If they lose engine power, they also lose some amount of control to keep the aircraft upright. Finally, all the testing takes place close to the ground where a failure of any component leaves little time to analyze the problem and take corrective action.

Ironically, the crashed XV-5A was repairable. After months of restoration and modification it reappeared as an XV-5B and was flown on more test missions.

The single Dassault Balzac killed two test pilots in two separate accidents. Ironically the Ryan XV-5As also killed two test pilots, but in two separate accidents.

Chapter 11

Metal Falling from the Sky

It had been a dreadful year at the flight test center with the loss of Maj. Carl Cross and Joe Walker in the XB-70/F-104 midair collision and Maj. Dave Tittle in the XV-5A crash in 1966. Every test pilot on the base was saddened by these losses. We hoped our luck would improve in 1967. Unfortunately, disaster soon struck again. On January 19, 1967, Dave Livingston and I were standing in the Test Operations scheduling room. It was a large room with ceiling-to-floor windows facing south, toward the main Edwards runway. A large board on the north wall of the room displayed the flight schedule for the day. It was shortly after 4:30 PM, about fifteen minutes before sunset, and we were checking the board to see what missions were scheduled for the following day.

Another pilot yelled out that he saw a fireball and smoke coming from the Mojave Desert west of the base. I stepped outside and saw a huge black cloud of smoke about three miles away. The wind was calm and visibility was at least fifty miles. It was a comfortable day in the middle of winter with a bright blue sky. The smoke appeared to be from an aircraft crash. Within moments we heard that an F-111A was "down," an expression pilots used for a crash.

Earlier in the day, Maj. Herbert "Herb" L. Brightwell, thirty-five, was scheduled to act as an instructor pilot in the ninth-built F-111A, serial number 63-9774. I had flown that particular aircraft many times and used it for wet runway tests the previous summer. Brightwell and I had gone through F-111A ground school together and checked out at the same time. He was a Tactical Air Command pilot who had flown the F-100 Super Sabre before coming to Edwards. Brightwell's student was Col. Donovan L. McCance, director of the TAC Category III test program. With McCance in the left seat and Brightwell in the right, they had taken off about 3 PM for a training flight. After performing maneuvers in the Edwards Restricted Area for an hour and forty-five minutes, they called the control tower for permission to make an instrument landing approach, followed by a touch-and-go landing on Runway 04.

An instrument landing approach was a unique procedure when done in the F-111A—I had made many of them. The right-seat pilot locked the General Electric air-to-ground radar onto the intersection of the main runway and the center taxiway, which was 7,500 feet from the approach end of Runway 04. The left-seat pilot set 2.5 degrees in the dive/climb digital counter. Then he followed the vertical and horizontal steering on his heads-up display and flew an approach similar to an instrument landing system (ILS) used by both military and civilian pilots even today. Unlike an ILS, the F-111A instrument landing approach used only onboard avionics and could be performed on any runway.

Colonel McCance descended to around 1,500 feet above the ground and turned the F-111A to a heading direct for Runway 04. About six miles from the airfield he lowered the landing gear. The F-111A was dual controlled, with a flight control stick and engine throttle available to each pilot. However, the left-seat pilot was the only person who could extend the landing gear and flaps.

The F-111A was the first Air Force operational sweep-wing aircraft—its wing position could be varied from a forward position of 16 degrees to 72.5 degrees aft. This allowed the aircraft to be flown slowly and still have plenty of lift at the 16-degree position for takeoff and landing. The 72.5-degree wing-sweep position was used for high-speed flight.

Logically, the cockpit-selector handle for moving the wing would be moved in the direction of wing motion, but a strange philosophy of cockpit design intruded. The first batch of F-111As had the wing moving in an opposite direction from the movement of the selector handle. Only one wing-sweep-selector handle was installed in the cockpit. It could only be operated from the left seat.

To make the wing go forward, the pilot moved the handle back. To make the wing go back, he moved the handle forward. The logic for this scheme was that the wing-selector handle should be considered as a throttle—to go fast, push the handle forward and the wing would go back, and to go slow, pull the handle back and the wing would go forward, slowing the aircraft. One can imagine how confusing its use was in actual operation. Nevertheless, the idea was incorporated—and on the first of these aircraft some enterprising test pilot doctored a decal near the wing position indicator that displayed *"Fore is aft and aft is fore."*

When I needed to sweep the wing, I simply moved the handle in either direction while watching the wing-sweep indicator—if the wing moved the wrong direction on the gauge, I would reverse the motion.

Colonel McCance set up a gradual descent and slowed the F-111A to approach speed. Because he had forgotten to lower the flaps and sweep the wings forward, the aircraft started to sink. In a moment of panic and without thinking, the colonel moved the wing-sweep handle forward, and the wings swept back. It was the opposite direction of what he desired. To regain control, he applied afterburner power, but the aircraft continued to sink and struck the ground about a mile and a half short of Runway 04. The aircraft hit with wings level in the desert and sheared the landing gear off.

Both pilots survived the crash landing. Major Brightwell opened his canopy, released his parachute fitting, and exited the damaged aircraft. He ran around the nose of the aircraft to assist Colonel McCance.

At that moment the aircraft caught fire and blew up because of fuel that had spilled on the ground. Brightwell sustained third-degree burns over half of his body. Colonel McCance was still sitting in the cockpit when the fire erupted and was not as badly burned.

Fortunately, two General Dynamic employees were driving nearby, departing the base at the end of their workday, and saw the accident. Donald N. Harris, chief of security, and James D. McMahon, chief of logistics, drove over the desert to the crash site and, despite the burning aircraft and exploding fuel cells, succeeded in removing the injured and dazed pilots to a place of safety.

Within minutes, an Edwards helicopter picked up the two burned pilots and flew them to the base hospital. Military doctors stabilized Brightwell and prepared him for an evacuation flight in a twin-engine C-131 aircraft. At 5 the next morning, Friday, January 20, the C-131 departed on the four-and-a-half-hour flight to the Burn Center at Brooks Air Force Base in San Antonio, Texas. An accident investigation board was convened to determine the cause of the mishap.

The next morning I arrived at Test Operations to find that I was scheduled to fly a dual-controlled F-104B on a night proficiency flight. Capt. Larry Curtis, whom Livingston and I had taught how to land sail, flew in my backseat. At sunset on a Friday evening, Curtis and I took off, and he practiced several instrument approaches to Runway 22. I took control of the aircraft when the Starfighter was about 200 feet above the ground and made several touch-and-go landings. As night came on we saw a group of floodlights illuminating the wreckage of Brightwell's F-111A, just a short distance beyond the far end of the runway we were using. It was a sobering reminder of the tragic accident the previous day. Curtis and I talked on the intercom about Brightwell and wondered how well he was recovering in Texas. We were particularly careful that night flying the F-104B. Between all the accidents that had already occurred in the Starfighter and Brightwell's crash, it was not a time to let our guard down.

The following day, January 21, Maj. Herb Brightwell died of burn injuries suffered in the crash.

The *Antelope Valley Press* in nearby Lancaster reported, "This was the first F-111 accident since the program began. The fleet of F-111s had logged 2,300 hours since it first flew on September 21, 1964. The Mach 2.5 fighter had been flown supersonic on more than 500 occasions and had been flown faster than Mach 2 a total of 127 times."[1]

A memorial service was held for Major Brightwell at the base chapel three days after he was killed. Following the death of an aviator, it was customary to fly a missing-man-formation flight at the conclusion of the service. The missing-man formation duplicated the four fingers of a person's right hand, and was even called a "fingertip" flight. To memorialize Brightwell, four F-104 Starfighters made a pass over the chapel about 1,000 feet above the ground and at around 300 knots. One Starfighter pulled up as the flight approached the chapel, signifying the missing man. Since I had known Brightwell in the F-111A Joint Test Force, I was asked to fly in the formation. It was an honor to be in the flight and pay tribute to his memory.

Later, the two General Dynamic employees were honored for risking their lives to rescue Brightwell and McCance. Col. Elmer W. Richardson, vice commander of the Air Force Flight Test Center, presented the Exceptional Service Medal, one of the highest honors for heroism that the Air Force can give to civilians, to Donald N. Harris and James D. McMahon. The citation accompanying the award praised the two for the exemplary courage and humanitarian regard that they displayed for their fellow man.

Another award ceremony was held in the office of the Flight Test Center commander when Maj. Gen. Hugh. B. Manson presented the nation's seventh highest award, the Airman's Medal (posthumously), to Major Brightwell's son, Kenneth. The citation that accompanied the Airman's Medal read:

> Major Herbert L. Brightwell distinguished himself by heroism involving voluntary risk of life at Edwards Air Force Base, California on 19 January 1967. On that date, Major Brightwell was participating as an Instructor Pilot in an aerial training flight when the aircraft crashed on landing. Major Brightwell, after successfully egressing from the burning aircraft and with complete disregard for his personal safety, remained at the aircraft to assist the other pilot. Despite the extreme hazard of exploding fuel cells, Major Brightwell was able to adequately instruct the other pilot and consequently saved his life at the sacrifice of his own. By his courageous action and humanitarian regard for his fellow man, Major Brightwell has reflected great credit upon himself and the United State Air Force.[2]

Test pilot Capt. Joseph F. "Joe" Stroface, a pilot assigned to the Fighter Branch, hit the jackpot in early 1967. He was notified that he had been selected to fly the "lifting body" class of aircraft. Stroface, thirty-five, graduated from the U.S. Military Academy at West Point in 1956. He flew the T-34 and T-28 in primary flying school at Marana, Arizona. Then, like me, he flew the T-33 at Webb Air Force Base in Texas. After graduating he was transferred to the 80th Fighter-Bomber Squadron at Itazuke Air Base in Japan. There he became combat ready in the F-100 Super Sabre.

Stroface told me of an unusual—and to him a slightly humorous—event that occurred flying the Super Sabre. He was tasked with flying an Air Force general in a tandem-seated F-100F model and taking him to a base where the general planned to give a speech. Stroface packed the general's uniform and presentation materials in a travel pod on the centerline of the aircraft. This attachment point usually carried weapons such as bombs, rockets, or napalm. Unfortunately, after takeoff a fire warning light illuminated, indicating that the Pratt & Whitney J57 jet engine was on fire. Following approved Air Force emergency procedures, Stroface jettisoned the travel pod—which reduces weight, giving the aircraft a better chance of reaching a runway—and safely landed the F-100F. The general's clothing and written speech fell from the aircraft and were never found. Stroface had followed military regulations, but the general was angry that he had lost his possessions and missed his scheduled talk. Stroface sympathized with the general's concern but privately got a chuckle out of the incident.

Joe Stroface graduated from the test pilot school six months before I did, and he and his classmate Capt. Jerry Gentry were assigned to the Fighter Branch. Stroface and I flew together in an F-104B dual-seated Starfighter several times, including one flight chasing the XB-70. When Mike Adams was selected to fly the X-15, Stroface took over for him on the Light Armed Reconnaissance Aircraft (LARA) program. LARA was based on a perceived need for a new type of "jungle fighting" versatile light-attack and observation aircraft. Existing aircraft (the Cessna O-1 Bird Dog and Cessna O-2 Skymaster) had too small a cargo capacity for this flexible role. The LARA specification called for a twin-engine two-man aircraft that could carry at least 2,400 pounds of cargo, six paratroops or stretchers, and be stressed for +8 and –3 Gs. It also had to fly at least 350 mph and take off in 800 feet. Various weapons had to be carried, including four 7.62-mm machine guns with two thousand rounds, and external weapons including a 20-mm gun pod and Sidewinder missiles.

Eleven proposals were submitted, and seven made the first cut: the Beech PD 183, Douglas D.855, General Dynamics/Convair Model 48 Charger, the Helio 1320, the Lockheed CL-760, a Martin design, and the North American/Rockwell NA300 (later designated the OV-10A). In August 1964, the NA300, by then nicknamed the Bronco, was selected for the program now called Counter-Insurgency (COIN). A contract for seven prototype aircraft was issued in October 1964.

General Dynamics/Convair protested the decision and built a prototype of the Model 48 Charger anyway and first flew it on November 29, 1964. This was also a twin-tail aircraft that had a similar layout to the Bronco. Joe Stroface made four test flights in the Model 48 and told me it was a very capable aircraft. The Charger, while capable of outperforming the OV-10A in some respects, crashed on October 19, 1965, after 196 test flights. Convair then dropped out of contention. North American Aviation test pilot Ed Gillespie made the first flight of the OV-10A on July 16, 1965, midway through the Charger's test program, and the Bronco became the premier COIN aircraft of the next thirty years.

After Stroface was notified he was being reassigned to the lifting-body program, Capt. Larry Curtis was chosen as his replacement. Stroface and Curtis flew on a commercial flight to the NAA plant at Port Columbus, Ohio, so that Curtis could get an introduction to the OV-10A aircraft and meet program officials.

Before traveling to Port Columbus for the test missions, Curtis sprained his ankle playing basketball at the base gym. His foot became swollen and he wrapped a bandage around it. The swelling prevented him from getting his foot into his flying boot. Curtis told me of his predicament and I offered to lend my boots to him since I had larger feet.

On February 1, 1967, Joe Stroface and Larry Curtis scheduled the last single-engine minimum-control speed test mission required by the test plan. They would fly a prototype OV-10A from the NAA facility at Port Columbus. Although the weather was marginal and North American officials recommended against the test, Stroface took off with Curtis in the backseat. The weather was overcast at 1,100 feet with visibility less than five miles in drizzle. The tests had to be accomplished below the clouds, extremely low altitude for single-engine tests. The pilots were anxious to finish the mission and return to Edwards. It was the setup for a classic case of "get-home-itis."

On board the Bronco were four 750-pound inert bombs and a full load of internal fuel. Stroface told me he was flying so low that he might not be able to jettison the bombs if trouble developed. Also, he did not trust the ejection seat on the Bronco.

"We were an accident ready to happen," Stroface told me when I interviewed him in 1995. Stroface explained that he planned to stay four to six miles from the nearby runway at Lockbourne Air Force Base (near Columbus, Ohio) to enable him to immediately land if the weather worsened. He slowed the Bronco to a speed just above the stall, with both the landing gear and flaps extended.

North American test pilots had previously determined the minimum control speeds of the Bronco. But Stroface and Curtis were looking for another characteristic in the OV-10A. According to Ed Gillespie, the chief OV-10 test pilot for North American Aviation whom I interviewed in 2006, their plan was to slow the aircraft down to just above the stall. And then Curtis was supposed to shut an engine down by pulling the condition lever to fuel shut-off without telling Stroface which engine he had shut down. Stroface would wait one second and then recover. Curtis would call the one-second delay to Stroface over the intercom.

Stroface and Curtis lost control of the OV-10A on that fateful day. One wing dropped and the OV-10A rolled into a steep dive. Neither pilot attempted to eject. Since the pilots started the tests at an extremely low altitude, they had only a few seconds before the Bronco crashed through a group of willow trees near a creek bed. A branch of a tree penetrated their cockpit and struck both pilots in the face and upper torso. Nearby a group of men were pruning an apple orchard. They rushed to the rescue of the two pilots still inside the wrecked Bronco. North American test personnel were monitoring the progress of the tests on a UHF radio and recognized the plane had gone down.

An Air Force emergency vehicle was dispatched to the scene of the accident, and the two pilots were in the Lockbourne base hospital within forty-five minutes. At the time, Larry Curtis was still alive. Later he was transferred to Ohio State University Hospital where he died of his massive injuries. Only a week earlier Curtis and I had flown a night proficiency flight in a F-104B. Now Curtis was dead. I was shocked and couldn't believe he was really gone. His death still troubles me, forty years later.

Joe Stroface suffered severe injuries to his face, eyes, and upper torso. He survived the accident but was medically disqualified from ever flying Air Force aircraft again. Stroface retired from the Air Force and became a civil service engineer at Edwards for many years. Eventually, he was able to qualify for an FAA pilot's license. One eye was rated at 20/200, making him blind from a medical point of view. The FAA allows one-eyed pilots to fly if they obtain a "statement of demonstrated ability" or waiver. This requires a flight with an FAA examiner who determines if the one-eyed pilot can operate an airplane with the normal level of safety. World War II pilot Saburo Sakai had only one eye. This disability didn't keep him from being Japan's highest ace to survive the war.

While I was stationed at Edwards, NASA and Air Force test pilots were flying other research aircraft besides the X-15 rocketship. A group of planes classified as wingless "lifting body" vehicles were being designed, flown, and landed on Rogers Dry Lake. The first vehicle was the Northrop M2-F1, a wooden 1,000-pound lightweight glider with fixed landing gear. It was towed above the lake bed behind a modified Pontiac automobile or carried aloft and released behind a Douglas C-47 Gooney Bird. The C-47, the military version of the legendary DC-3, had been an excellent glider tug during World War II in such campaigns as Sicily and Normandy.

The other lifting-body vehicles were the Northrop M2-F2 and HL-10. Both were all-metal aircraft with a retractable landing gear and rocket motors. They would be attached to a pylon under the wing of a B-52 and flown to altitude just like the X-15. Each of the Northrop metal gliders cost $1.2 million; they were low-lift, high-drag one-of-a-kind gliding vehicles designed to help engineers determine how the yet-to-be-designed space shuttle would land after returning from earth orbit.

Bruce A. Peterson, thirty-four, was a NASA test pilot stationed at the Dryden Research Center. Peterson was born in Washburn, North Dakota. He joined the U.S. Marine Corps and graduated from flight school in 1953. While he was a squadron pilot he flew the propeller-powered F6F Hellcat and AD Skyraider and the turbine-powered F2H Banshee and F9F Cougar.

Peterson told me in 2005 that he had a close call flying the twin-engine F2H Banshee from Marine Corps Air Station Kaneohe on the north side of Oahu. One of his engines caught fire during flight. The engine eventually blew up and smoke was streaming from the plane. With a full load of fuel, Peterson landed and used the arresting gear on the far end of the runway to stop. He opened the canopy and tried to evacuate the burning plane as fast as possible. Peterson forgot to unhook his Mae West life preserver from the kit that held his life raft.

"I was jerked back toward the cockpit when I tried to get out," Peterson explained to me.

Anxious to extinguish the flames, firemen hit the plane with a large amount of foam. Unfortunately Peterson was still struggling to get released from his equipment.

"Foam hit me square in the chest," Peterson told me. "The force of the foam knocked me over, and I ended up spread-eagle in the cockpit."

A Marine flight surgeon also responded to Peterson's in-flight emergency. When he was safely removed from the Banshee, the two of them walked over to the Officers Club. Thankful Peterson was not injured, they quickly downed a couple of drinks. While drinking Peterson got his foot caught between his bar stool and the foot rail and fell off his seat. He ended up on the floor and hurt his right arm.

The flight surgeon looked at Peterson's arm and said, "I think it's broken," adding, "I can't set it until tomorrow morning." Peterson didn't explain why—maybe because the doctor had been drinking.

The two of them continued drinking long into the night—Bruce Peterson simply used his left arm. Only a Marine would save a disabled airplane but get injured in a drinking celebration!

Bruce Peterson separated from the Marine Corps and attended college at California State Polytechnic College at San Luis Obispo, where he received his bachelor of science degree in aeronautical engineering in 1960. He then joined NASA at Dryden Research Center as an engineer. In 1962 he attended the Air Force Test Pilot School in a class with the three Air Force captains, Mike Adams, Bob Parsons, and Dave Scott. After graduating from the school, Peterson returned to NASA and started flying test missions in the wooden M2-F1 glider.

On one flight, Peterson landed with sufficient force to shear off the landing wheels, and the M2-F1 sustained minor damage. During the tow to altitude, the automobile-type shock absorbers had become chilled and the cold hydraulic fluid simply failed to function on touchdown. NASA replaced the Cessna 150 landing gear installed on the M2-F1 with the more rugged gear from a Cessna 180.

After Capt. Joe Stroface was injured in the crash of the OV-10A, Capt. Jerauld R. "Jerry" Gentry was assigned as the Air Force lifting-body pilot. Gentry, thirty, was from Enid, Oklahoma. He graduated from Kemper Military School and then the U.S. Naval Academy in 1957. Both Gentry and Stroface were in the test pilot school class that preceded mine. Later Gentry became involved in some extremely hazardous rolling maneuvers on an M2-F1 flight.

Earlier in the year I was with Gentry when he borrowed a camera from the photographer who had just flown with him in the backseat of his T-38. Gentry and I had just returned from separate flights and were sitting in a bus used to transport pilots from Test Operations to the fighter flight line. Without thinking, Gentry picked up the camera and pointed it out the window of the bus to check the viewfinder. Unfortunately, the bus was passing six F-106 Delta Dart interceptors that were standing nuclear alert on the Edwards ramp. The Air Police guarding the interceptors stopped the bus and challenged Gentry as to why he was taking photos. Gentry talked his way out of the predicament and we were allowed to proceed.

Gentry couldn't talk his way out of a close call in the M2-F1. He was on takeoff behind the C-47 tow ship when he encountered severe propwash.

Turbulence from the C-47's two engines affected the M2-F1 with its short wingspan. His plane went into lateral oscillations that became larger on every cycle. Eventually the M2-F1 was upside down just 400 feet above the lake bed when Gentry cut loose from the tow aircraft. He completed a barrel roll, leveled the wings, and landed. It was a breathtaking experience for other pilots and engineers who watched the unexpected acrobatics. After NASA management heard about the close call, they closed the M2-F1 flight test program. It had served its purpose. It proved that a lifting-body shape could fly and encouraged further research with supersonic, rocket-powered lifting bodies, to determine if the shapes so desirable for hypersonic flight could safely fly from supersonic speeds down to landing. When the tubby M2-F1 completed its last air tow in August 1966 work was already well along on the heavyweight aluminum follow-on aircraft—the M2-F2 and HL-10.

The maiden flight of the M2-F2 was made just before the last flight of the M2-F1. During the brief flight—not quite four minutes—NASA test pilot Milt Thompson, who had flown the X-15, dropped away from the B-52 mothership at 42,000 feet. He made a practice landing at 23,000 feet, turned 180 degrees to the left, lowered the landing gear, and touched down exactly at the planned aiming point on Rogers Dry Lake. The first flight had been an unqualified success. Five months later, Thompson, Peterson, and Gentry had completed an additional thirteen flights in the craft. Following the last flight, NASA brought the craft into the hangar for installation of its XLR-11 rocket engine.

While the M2-F2 was undergoing glide flights, the HL-10 was being prepared for its first flight. The HL-10 (for horizontal landing, tenth shape designed) had a flat bottom and a rounded top, unlike the M2-F2, which had a cone-shaped underside. The HL-10 had three vertical fins and a delta platform. Both the M2-F2 and HL-10 weighed about 4,600 pounds and used many off-the-shelf components from the T-38, T-39, and F-106.

According to the NASA simulator, the HL-10 was more stable and would have much better handling and lift-to-drag ratio than the M2-F2. The pilots who flew the simulator were suspicious and skeptical of the results, saying it was too good to be true.

On December 22, 1966, Bruce Peterson climbed into the HL-10 for its first flight. The B-52 taxied to the main Edwards runway and took off. The launch point was about three miles east of Rogers Dry Lake abeam lake bed Runway 18, the same runway used by the X-15. The HL-10 was released on a north heading, and Peterson planned to make two 90-degree left turns, the same flight path used for the first M2-F2 flight. Peterson was launched from the B-52 at 45,000 feet and 170 knots. He immediately became aware of a high-frequency buffet in pitch and roll. As the plane descended and picked up speed, the buffet became worse. The metal HL-10 was too responsive in pitch and unresponsive in roll. So Peterson kept the angle of attack low, increased speed, and touched down on the lake bed at 280 knots, about 30 knots faster than planned. The poor control surprised the NASA engineers, who went back to their drawing boards.

The M2-F2 spent five months in the hangar for installation and ground tests of the XLR-11 rocket engine. Two glide flights were planned to test the handling of the vehicle with the added weight of the rocket engine and the modifications needed to fit it in the vehicle.

Jerry Gentry made the M2-F2's first glide flight with the engine installed on May 2, 1967. It was his fifth flight in the vehicle. Gentry was no big fan of the M2-F2's handling qualities, and this flight intensified his dislike. Several times during the descent Gentry thought he was on the verge of losing control of the glider. On final approach, he again felt he was not in control. The landing was successful despite the problems. At the post flight debriefing, Gentry said he did not want to fly the M2-F2 again until it was fixed. The decision was made to have Peterson fly it and see whether he thought it was really that bad.

Bruce Peterson made the next flight in the M2-F2 on May 10, 1967. The flight was normal from launch to the final turn toward the lake bed. Research maneuvers were conducted on the first two legs of the U-shaped flight path. A wind gust hit the glider, causing it to oscillate in bank, but Peterson used rudder to control it. The rudder aggravated the Dutch roll motions, and the aircraft made two turns to a bank angle of 140 degrees at a roll rate of 200 degrees per second. With the momentary loss of control, Peterson was off heading for a landing on lake-bed Runway 18, and there was no time left to make any changes before making the landing flare. As Peterson began the flare, he saw a rescue helicopter directly in front of him. He made three radio calls, all concerning the nearness of the helicopter and the possibility of a midair collision. The M2-F2 cleared the helicopter, but as the flare was completed the glider hit the lake bed with its landing gear still retracted. The glider bounced back in the air touching down again about 80 feet later with the landing gear partly extended. After skidding a short distance, the M2-F2 turned sideways and began rolling. The glider made six rolling revolutions and finally slammed into the lake bed upside down. The last thing Peterson remembered before blacking out was a high deceleration and the plane starting to roll at an increasing rate.

A helicopter was the first to reach the M2-F2. A NASA crewman jumped out and ran over to the plane to help Peterson. When the crewman crawled under the wreckage, he found Peterson hanging upside down in his straps with severe facial injuries. Peterson was pulled clear of the demolished glider, put on a stretcher, and loaded onto the helicopter for transport to the base hospital.

Surprisingly, Peterson only had minor physical injuries. He had broken a bone in his right hand and had lost skin from his forehead and right eye area. He was instrumented on this flight, and his body acceleration recorded. The data revealed that he was subjected to less than 5 Gs throughout the entire crash sequence. Most of the vehicle's energy was dissipated in the tumbling and rolling motions. Peterson might have survived the accident injury-free if the glider had not ended on its canopy on its last tumble.

The crash of the M2-F2 was recorded by NASA engineers and replayed over and over for test pilots at Edwards to watch. I thought it was incredible that my friend Bruce Peterson had survived the accident. We all hoped and prayed he would recover from his injuries and fly again.

After Peterson was given emergency treatment in the Edwards hospital, he was moved to March Air Force Base in Riverside, California, because of the damage to the area around his eye. His eyelid had been torn away. He was finally transferred to the University of California in Los Angeles for specialized treatment for his eye injury. The eyeball itself was not seriously damaged. Peterson could see, but the loss of the eyelid would ultimately result in damage to the eyeball unless an eyelid could be reconstructed. Peterson told me that the doctors needed skin from him to construct a new upper eyelid. The best skin for this purpose was the foreskin. They asked Peterson if he had been circumcised. Peterson told me he had been circumcised and was happy the doctors did not graft skin from that area to his eye. In a humorous vein he said that if foreskin had been grafted to his upper eyelid, every time he saw a pretty girl the eyelid would probably go into an uncontrollable flutter!

During the lengthy process of healing and preparation to repair the damage, Peterson contracted a staph infection, which eventually resulted in loss of vision in that eye a month or so after the crash. He was grounded by NASA and wore a black plastic eye patch the rest of his life.

Bruce Peterson's crash in the M2-F2 on May 10, 1967, wasn't the only accident that day. Outside of Dallas, Texas, 1,200 miles east of Edwards, another tragedy unfolded. Civilian test pilot Stuart G. "Stu" Madison and two of his crew members were killed in the LTV-Hiller-Ryan XC-142A. Madison had tested the XC-142A at Edwards in the previous months.

The first vertical takeoff and landing concepts to be tried in the 1950s were three "tail sitting" airplanes, the Lockheed XFV-1 Salmon, the Convair XFY-1 Pogo, and the Ryan X-13 Vertijet. It was extremely difficult for the pilot to take off and land in these aircraft while lying on his back, so another concept was developed. The next design kept the main body of the aircraft level with the runway but tilted the wings or engines from a horizontal to a vertical plane.

In the 1960s, the Army, Navy, and Air Force began work on the development of a prototype VTOL airplane that could augment helicopters in transport missions. In the case of the XC-142A, the entire wing and its attached four engines rotated straight up. Tilting the wing and engines skyward permitted vertical takeoff like a helicopter, and then the wing and engines were gradually tilted forward to provide the greater speed of a fixed-wing aircraft in forward flight. The plane used four cross-linked 3,080 shaft-horsepower General Electric T64-GE-1 engines, each driving a 15.5-foot four-bladed propeller. The engines were linked together so that a single engine could turn all four propellers and the tail rotor. Roll was controlled by differential propeller pitch, and pitch by an eight-foot, three-bladed variable pitch tail rotor. Ailerons powered by propeller slipstream provided yaw. The wing tilted through 100 degrees, allowing the XC-142A to hover in a tailwind. The tail rotor folded to the port side to reduce the stowage length and to protect it against accidental damage during loading. The cargo aircraft was 58 feet long and had a wingspan of 67 feet. In tests the XC-142A was flown from airspeeds of 35 mph backward to 400 mph forward.

Aside from the pivotal wing, however, the XC-142A looked like a somewhat stubby, small cargo hauler. Its designers had provided it with a compartment capable of carrying thirty-two fully equipped combat troops or 8,000 pounds of cargo in a vertical-lift mode. Top speed in horizontal flight was 430 mph, and, fitted with special fuel tanks, it would have an ocean-hopping ferry range of some 3,800 miles. Some enthusiasts even predicted a submarine-hunting role, with the radical aircraft operating from Navy antisubmarine task forces. The end product of this thinking was the world's largest VTOL aircraft and the first such American vehicle with enough payload capability to permit operational evaluation by the military services.

The XC-142 Tri-Service Test Force of 150 military and civilian personnel was established at the Air Force Flight Test Center, and it eventually demonstrated the test center's ability to handle a difficult and innovative evaluation program. Lt. Col. Jesse P. Jacobs directed the test force, which included Army, Navy, and Marine Corps as well as Air Force pilots. The first of five XC-142As built by LTV Aerospace arrived at Edwards on July 9, 1965, following a flight from Dallas—the first long-distance hop of a VTOL aircraft. The new transport proved easy to convert from vertical to horizontal flight, and the test force gave it high marks for its speed and stability in hover. Over the following months the team conducted a wide variety of tests.

Since all test flights at Edwards required a safety chase, I flew a loose formation position on the XC-142A several times in a T-33 or T-38. Every time I chased the plane it was in horizontal flight with the wings level. Nevertheless, the XC-142A was a strange-looking bird and, like the XV-6 Kestrel, French Balzac, and Ryan XV-5 Vertifan, suffered many mishaps. I had to be very vigilant every moment I accompanied this unique flying machine.

Of the five XC-142A aircraft built and tested at Edwards, one experienced a ground loop, causing extensive damage to the wing and propeller. One pilot made a hard landing in the vertical mode; one aircraft sustained severe damage to the wing, aileron, and number-two nacelle from an engine turbine failure; and one experienced a ground accident caused by pilot error when he failed to activate the hydraulic system, which resulted in no brakes or nosewheel steering.

The most serious of the mishaps resulted from a tail rotor drive-shaft failure and caused three fatalities on May 10, 1967. The accident occurred to XC-142A serial number 62-6921 and took place near the LTV (Ling-Temco-Vought) plant in Dallas, Texas. Stu Madison placed the ill-fated prototype in a rapid dive from 8,000 feet to 3,000 feet, effectively simulating a pilot rescue under combat conditions. When the driveshaft failed the XC-142A nosed over at low altitude, and the crew could not recover. The crash occurred in a heavily wooded area where fire started after impact. All three crew members were killed.

For all of its great promise, however, the four-engine VTOL transport was still somewhat ahead of its time and never reached its full potential. The airplane was underpowered and was subject to a number of handling difficulties including excessive vibration and cockpit noise resulting in pilot fatigue. Soon attrition left only a single aircraft to continue the trials, and the XC-142A program was eventually canceled.

While the XC-142A program lasted only a few years, the X-15 program was in its eighth year of successful data gathering. Among those who flew the X-15 was Maj. William J. "Pete" Knight. He was born in Noblesville, Indiana. Knight, thirty-eight, whose Depression-era family home had no indoor plumbing, spent summers working on his grandparents' farm. He earned the nickname Pete in grade school when he wiggled his nose like Peter Rabbit to get a girl's attention. Knight attended Butler and Purdue universities before joining the Aviation Cadet Program.

"I became interested in flying in high school," Knight said. "But it was never a burning desire."

Nevertheless, Knight launched himself on what would become a long and storied flying career, earning him the description by his peers of being unflappable. Flying an F-89D for the 438th Fighter Interceptor Squadron, he won the prestigious Allison Jet Trophy Race in September 1954. After completing his undergraduate education with a degree in aeronautical engineering from the Air Force Institute of Technology in 1958, he attended the test pilot school at Edwards where he graduated later that same year. He remained at Edwards where he served as project test pilot on the F-100, F-101, and F-104 and later the T-38 and F-5 test programs. In 1960, he was one of six test pilots selected to fly the X-20 Dyna-Soar, which was slated to become the first winged orbital space vehicle capable of lifting reentries and conventional landings. After the X-20 program was canceled in 1963, he completed the pre-astronaut training curriculum at the new Aerospace Research Pilot School at Edwards in the summer of 1964 and was selected to fly the X-15 in 1965. Pete Knight and I flew together frequently. He checked me out as an instructor pilot in the T-38.

Six weeks after Stu Madison's fatal accident in Texas, near disaster occurred to Pete Knight when he was flying the X-15 rocketship. He was a pilot who had more than his share of eventful flights in the airplane. While the aircraft was climbing through 107,000 feet at Mach 4.17 on June 29, 1967, it suffered a total electrical failure, and most onboard systems shut down. After arching over at 173,000 feet, Knight calmly set up a visual approach, and, resorting to old-fashioned seat-of-the-pants flying, he glided down to a safe emergency landing at Mud Dry Lake in Nevada, which was one of several landing areas to be used in case of a problem.

Knight recalled, "Anybody who says they are not scared is either lying or they are not too bright. You're scared, but you control your fear."

For his remarkable feat of airmanship that day, he was awarded a Distinguished Flying Cross.

As it tuned out, Mud Dry Lake also had been used earlier for an emergency landing. On November 9, 1962, NASA test pilot John B. "Jack" McKay crashed the number two X-15 there. McKay, forty, was born in Portsmouth, Virginia, and had been a Navy pilot in World War II. He received a bachelor of science degree from Virginia Polytechnic Institute in 1950. McKay then joined NACA in 1951 as a Douglas D-558-II and Bell X-1 rocket test pilot. He was also a project pilot on F-100, F-102, F-104, and F-107 test programs before being assigned as third NASA X-15 pilot on October 25, 1961.

McKay was severely injured in the crash, and the aircraft was substantially damaged. Instead of just repairing the aircraft, program officials elected to take this opportunity to modify the aircraft for scramjet engine tests. Approximately a year after the accident, the number two X-15, now named the X-15A-2, emerged like a phoenix from the bowels of the North American Aviation production facility on the south side of Los Angeles International Airport. After the accident the X-15 had been extensively modified to carry an experimental scramjet engine up to speeds of over 5,000 mph. Modifications included the addition of a twenty-eight-inch plug in the middle of the fuselage between the liquid oxygen (LOX) and ammonia tanks. A spherical tank was installed in this extra space to carry the liquid hydrogen fuel for the scramjet engine, which was to be carried on the lower ventral fin stub.

Another major modification was the addition of two huge external tanks for additional main engine propellants. One tank carried LOX and the other liquid ammonia. Together these tanks carried an additional eighteen hundred gallons of propellants that corresponded to about sixty seconds of engine burn time. The total propellant was almost double that carried by the standard X-15. One might assume that this would provide a big increase in maximum speed. Realistically, the actual speed increase was disappointingly small. The extra weight and tremendous increase in drag due to these tanks severely reduced the rate of rotation to achieve the desired climb angle and also reduced the rate of climb and acceleration. With full external tanks, the aircraft weighed almost 57,000 pounds compared to 33,000 pounds for the standard, fully fueled aircraft. All of the external fuel would be used up just to reach Mach 2 at 70,000 feet, where the tanks were jettisoned. The internal fuel did not provide the same total acceleration as the standard airplane, since the new airplane was heavier and had somewhat more drag, particularly with its ablative coating.

In addition, the basic aircraft structure had to be protected from the increased aerodynamic heating predicted at the higher planned speeds. The proposed solution was to cover the entire airplane with an ablative material developed to protect missile nose cones. This material acted as an insulator. More effectively, it acted as a dissipater of heat through a burning process that produced a heat resistant char layer. Variations of this ablative material were used to protect the Mercury, Gemini, and Apollo spacecrafts. This material was installed on the X-15 in both premolded pieces and as a spray-on coating, and in one case with a putty knife. The material was pink—which did not quite fit the macho image of an Edwards test pilot.

Pete Knight allegedly said, "There's no way in hell I'm going to fly a pink airplane!"

Luckily, an additional protective coating was required over the ablative material due to its sensitivity to liquid oxygen. This protective coating was white—an acceptable color.

Once X-15A-2 was modified, NASA and the Air Force began a flight envelope-expansion program on the basic airplane, without tanks, to determine the effects of the basic airplane modification on the aircraft's stability, control, and handling qualities. The effects were not anticipated to be significant because the external configuration of the modified basic aircraft had not changed much. However, program engineers needed to accurately quantify the aircraft's flying characteristics before the external tanks and scramjet engine were installed.

Maj. Robert A. "Bob" Rushworth was scheduled to fly the first envelope-expansion flight. Bob was a good friend—we had flown several times together in the T-38 and he had checked me out as an instructor pilot in the F-104.

Bob Rushworth, forty-three, was born in Madison, Maine, and earned his pilot's wings during World War II. He flew C-46 and C-47 aircraft while assigned to the 12th Combat Cargo Squadron. After five years with the Air Force Reserve and Air National Guard, Rushworth was recalled to active duty in February 1951 during the Korean War. His assignment was as an F-80C Shooting Star pilot with the 49th Fighter Interceptor Squadron at Dow Air Force Base in his home state of Maine. Following graduation from the Air Force Institute of Technology, Rushworth stayed at Wright-Patterson Air Force Base. He served at the Directorate of Flight and All-Weather Testing, where he specialized in the development and flight testing of experimental automatic flight control systems. In 1956, he began a tour at Edwards to attend the test pilot school. After graduation in 1957 he was assigned to the Air Force Flight Test Center, first as an experimental flight test officer and eventually as assistant director, Flight Test Operations. During this period, Rushworth test flew the F-101, F-104, and other jet fighters, as well as the X-15. By the time of the first envelope-expansion mission in the X-15A-2, Rushworth had flown twenty-one flights in the rocketship.

In his first flight in the X-15A-2 Rushworth flew to a maximum speed of Mach 4.59. Everything worked fine and the airplane flew reasonably well. He also flew the second flight and achieved a maximum Mach of 5.23. But as he was decelerating back through about 4.5 Mach, he heard a tremendous bang under his feet, almost like an explosion. After he descended, Maj. Don Sorlie, his safety chase pilot, confirmed that the nose gear had extended, but he had to move in closely to determine the extent of the damage. Sorlie informed Rushworth that the tires were badly burned, but they appeared to be intact. The nose gear strut, door, and wheel well appeared to be charred but otherwise appeared normal. The tires failed at touchdown and rapidly disintegrated during the high-speed rollout, but the gear survived. From the flight data, NASA learned that the nose gear had accidentally deployed at Mach 4.4. The temperature created by the impact of the air at that speed was over

1,000 degrees Fahrenheit. That obviously accounted for the burned tires and charred nose gear.

During the rollout after landing, Rushworth was heard to comment sarcastically, "Thank you, Mr. North American."

The next flight was scheduled about six weeks later, and Rushworth was again scheduled to fly it. The flight plan called for a flight up to Mach 5. Additional stability and control data was to be obtained at roughly 100,000 feet. The flight proceeded as planned up through another burnout. As the aircraft decelerated through Mach 4.5, Rushworth heard another bang and the aircraft pitched down and began to oscillate in roll and yaw. Again, he was not certain of what had happened, but he suspected the landing gear system had malfunctioned again. The bang was not as loud as on the previous flight, so he assumed that it might just be the nose gear scoop door opening rather than the entire gear extending. He advised the control room of his suspicions, but again the control room could not identify the problem nor really help him other than to offer sympathy. Rushworth had to wrestle with the airplane to regain complete control and fly it on to Edwards. On arrival at Edwards, Maj. Joe Engle joined up and confirmed that the nose gear scoop door was open. As Rushworth came level following the landing flare, he pulled the gear handle and nothing happened. He immediately pulled the handle again and heard the nose gear deploy. Engle noted that the nose gear was extending very slowly.

Engle called out, "Hold it off," to allow the nose gear time to extend and lock.

Rushworth managed to hold the airplane in the air just long enough to let the nose gear lock before the main skids touched the lake bed. The nose slammed down hard, but the nose tires stayed intact for the first thousand feet of the rollout and then disintegrated. Extensive heat damage was determined by visual inspection of the nose gear and the nose-gear wheel well. The scoop door opening allowed hot air to enter the nose gear compartment, which burned the tires, melted aluminum tubing, and caused the nose gear to bind up during extension. This incident had the potential to be more catastrophic than the previous one. Luckily, it was not. Rushworth was beginning to have mixed emotions about the joys of flying the X-15A-2. He was getting upset with these surprises and was also getting gun shy. What might happen the next time?

Rushworth's next flight, in February 1965, proceeded normally until the aircraft decelerated through Mach 4.4. At that time, he felt, rather than heard, something let go. This time he could not deduce what happened. He had to wait until Joe Engle joined up to be informed that his right main skid was down. This caused some concern that the skid might not function properly on touchdown and would result in a failure of the gear itself. Rushworth made a very smooth landing and the gear remained intact.

During the rollout, Rushworth shouted, "Boy, I'll tell you, I've had enough of this!"

When Rushworth finally got out of the airplane, he turned around and kicked it. This was not the first time Rushworth was displeased with an aircraft. Our F-104A, serial number 56-0748, assigned to the Fighter Branch was very underpowered. The plane had been modified with a very accurate airspeed system and used to pace other test aircraft and calibrate their pitot-static system. All of us flew the plane from time to time and were aware of its low performance. The maintenance department had changed the General Electric J79 engine several times without any improvement.

One time when Rushworth was returning from a chase mission, he entered the Edwards pattern and lowered the landing gear. On base turn he selected full flaps and the Starfighter started to sink. To keep the F-104 from crashing he lit the afterburner, rolled out on final, and landed. To show his irritation, he taxied back to the parking position on the ramp, swung the tail of the Starfighter so as to point the exhaust directly into the hangar, and ran the engine up to full military power. This action caused havoc in the hangar. Paperwork and parts were scattered all over, but the maintenance supervisors got his message.

Postflight examination of Rushworth's flight in the X-15A-2 revealed a bent right-hand gear up-lock hook. Further analysis indicated that aerodynamic heating caused the new longer main gear strut to bow outward more than normal, which failed the up-lock hook and allowed the gear to deploy. The up-lock hook was subsequently modified to compensate for the extra deflection of the longer strut.

The X-15 pilots had no more landing gear incidents because of that modification, but they did put a small humorous sign above the landing gear T-handle that read *"Do not extend landing gear above Mach 4."*

Rushworth made the first flight of the X-15A-2 with external fuel tanks in November 1965. The intent of this flight was to demonstrate jettison of empty tanks at the design jettison conditions of Mach 2 at 70,000 feet. Fortunately the flight went as planned. The tanks separated cleanly and Rushworth made an uneventful recovery at Edwards.

Rushworth's next flight was the first flight with fully loaded external tanks. The major concern on this flight was the possibility of an unplanned partially full tank jettison due to an emergency. As a person might guess, with all the incidents that had already occurred, on this flight the tanks had to be jettisoned in a partially full condition. Shortly after X-15A-2 launch and engine light, Rushworth could not confirm that he was using fuel out of the external tank. His propellant flow meters indicated that he was using LOX out of the external tank, but no fuel. If this were indeed true, he would soon have an uncontrollable airplane with a large asymmetry in internal and external fuel tank loading. He jettisoned the tanks at 1.6 Mach at 41,000 feet and then jettisoned his internal fuel and made an emergency landing at Mud Lake. Capt. Larry Curtis flew as a safety chase for Rushworth and helped him safely land on the lake bed. This mission was Rushworth's last scheduled X-15 flight. He told one of the other X-15 pilots he was glad to finally leave the program. Sarcastically, he said he wanted to try something less hazardous for a while, like going into combat in Vietnam!

Pete Knight inherited the envelope-expansion program from Bob Rushworth and flew all the remaining flights in the X-15A-2 airplane. Knight's fifth flight in the X-15 was his first flight in the X-15A-2, which now carried external fuel tanks. The flight was launched and flown almost exactly as planned. Knight reached a maximum Mach number of 6.33 at 97,000 feet, a world speed record at the time. Tank jettison was clean and both recovery chutes deployed.

Due to wet lake beds resulting from winter rain, the next flight did not occur until the following May, nearly six months later. The flight was flown as planned with a maximum Mach number of 4.75 and maximum altitude of 97,600 feet, slightly higher than planned. All the flight objectives were achieved. The scramjet had no significant effect on aircraft handling qualities. One significant result of this flight was the complete erosion of the ablative material from the leading edge of the ventral behind the spike of the dummy scramjet. Maj. Mike Adams, who had now been on the X-15 program for a year, assisted Knight as he landed on Rogers Dry Lake.

Three months later Knight made a flight with a complete ablative coating on the aircraft and the dummy scramjet. He achieved Mach 4.94 at an altitude of 85,000 feet. Postflight investigation revealed severe erosion of the protective ablative coating.

Although not planned as such, the next flight on October 3, 1967, would turn out to be the last flight of the X-15A-2. It was planned to be a high-speed flight to Mach 6.5. The ablative coating applied for the previous flight was refurbished as necessary and the aircraft prepared for flight. This was the first flight with all of the elements needed for a high-speed flight. The configuration included the ablative coating, a canopy eyelid (a device that covered the entire windscreen during the high-speed flight but could be opened by the pilot for landing), the dummy scramjet, and full external fuel tanks.

Knight's climb went according to plan, and at sixty seconds he prepared to jettison the tanks. The tanks came off at sixty-seven seconds, and, according to Knight, they came off hard, very hard. He maintained that angle of attack until he came level at approximately 100,000 feet. Knight continued to accelerate in level flight until he reached an indicated 6,500 feet per second, or roughly Mach 6.5. At this time, 141 seconds after the launch, he shut the engine down. The thrust seemed to fall off slowly, and velocity increased to 6,630 feet per second, or Mach 6.7, another world speed record and the fastest speed ever flown by a winged vehicle. It's a record that stands to this day. Unknown to Knight, the intense heat from the shock waves off the dummy scramjet was burning off the ablative coating and melting huge holes in the skin of the ventral. He was high on energy, unable to jettison propellants, and unsure of the condition of the aircraft. He finally got the airplane turned around over the south end of the lake bed but was still supersonic at an altitude of 40,000 feet. As he rolled out heading northbound, the dummy scramjet tore off the airplane and tumbled on down to impact on the bombing range without a recovery chute, but Knight was able to make a successful landing.

Examination of the airplane after the flight revealed extensive damage to the ventral and lower fuselage structure. The tough nickel-steel skin was melted through like butter, which created several large holes in the ventral. The recovered dummy scramjet showed similar heating damage. Everything inside the ventral was burned, including the hydraulic lines that connected to the speed brake actuators. These lines had not burned through. If they had, NASA would have lost the aircraft, and possibly Knight. The damage to the aircraft was a sobering sight. But Pete Knight had accomplished a major milestone—he attained a speed that remains to this day the highest ever flown in an airplane.

The airplane was returned to North American Aviation for repairs. The near loss of X-15A-2 precipitated a reassessment of the overall X-15 program. The record-breaking aircraft never flew again. It was repaired and then sent to the Air Force Museum at Wright-Patterson Air Force Base in Dayton, Ohio, where it is still proudly displayed.

During his sixteen flights in the rocket-powered craft, Knight also became one of only eight pilots to earn astronaut's wings by flying an airplane into space when he climbed to 280,500 feet (more than fifty-three miles above the earth's surface) on October 17, 1967.

Back in October 1965 Maj. Mike Adams was one of eight military test pilots selected for the Air Force Manned Orbiting Laboratory (MOL) program. Two of my classmates from the school, Dick Truly and Jack Finley, both Navy pilots, were also selected. Capt. Jim Taylor, the unofficial winner of our Fleet enema contest back at Brooks, when we competed for the title of best physical specimen, was also selected. If these MOL astronaut candidates were required to take another enema, I'm sure Taylor won that contest again. The MOL program would be the military's first manned space program. Secretary of Defense Robert McNamara had canceled Dyna-Soar, the other military manned space program, in 1963.

After a while Mike Adams became dissatisfied with the slow progress being made to launch the MOL program into space and applied for an opening in the joint USAF and NASA X-15 program in early summer of 1966. He made the right call—the MOL program was canceled in 1969.

Adams was accepted for the X-15 position and returned to the Fighter Branch at Test Operations later in the summer of 1966. He was happy to be back at Edwards, and we were all pleased to see him return after only nine months away on the MOL program.

His first two attempts to fly the X-15 were aborted for aircraft problems while airborne and still hooked up under the wing of the Boeing B-52. Adams' bad luck had not changed.

He made his first actual X-15 flight on October 6, 1966, in the number one aircraft. The test plan called for him to shut down the rocket engine after 129 seconds of burn. A fuel tank bulkhead failure caused premature engine shutdown ninety seconds after launch, however. After an unsuccessful attempt to relight the engine, ground controller Pete Knight directed Adams to make an emergency landing on Cuddeback Dry Lake, about forty-three miles northeast of his intended landing on Rogers Dry Lake.

Adams responded with the comment, "Wonder why I picked a shutdown?"

The last thing any pilot needed on his first flight in a totally different aircraft was an emergency. His comment was low key, but it also expressed his concern that it was he who had experienced the shutdown, rather than one of the experienced X-15 pilots in the program.

Adams arrived at high key over Cuddeback with plenty of altitude to conduct the emergency landing and commented to Knight, "This thing is sort of fun to fly."

Knight was not in the mood for frivolity in the middle of an emergency and said, "Well, let's talk about it after you get it on the ground."

After he returned to Edwards and debriefed his flight, Adams flew an afternoon mission in the T-38 at Test Operations, as scheduled. For the second time in one day, Adams had an inadvertent engine shutdown. The T-38 had two engines, unlike the single rocket engine in the X-15, so he brought the plane back safely on one engine to the Edwards main runway. He was using up his nine lives fast.

Adams' second X-15 flight was in the number three aircraft. On this flight everything proceeded as planned, except he lacked UHF radio contact with the ground controller from prelaunch to overflight at Cuddeback Dry Lake.

During his third flight, now in the number one aircraft, Adams lost cabin pressure while climbing through 77,000 feet, causing his pressure suit to inflate. (Fortunately, his copilot dog, Tripod, was not with him on this flight.) Flying in a pressure suit made it more difficult to fly, but it was only the beginning of his troubles. As Adams flew the X-15 through a peak altitude of 133,000 feet, his inertial computer failed, causing a loss of all velocity, altitude, and climb-rate indications. Even without this information, Adams made a successful reentry and return to Edwards.

On approach, he radioed to Pete Knight, "I thought you said every once in a while something goes wrong?"

Mike Adams had certainly experienced more than his share of aircraft failures. During the postflight debriefing, he indicated that he had suffered from vertigo during the climb out.

Adams commented, "I did not know what the hell I was doing."

This was not a good sign for a test pilot who was rapidly burning up his nine lives. This problem would return later and played a crucial factor in his seventh flight.

On his fourth flight, Adams had problems controlling his climb angle due to an apparent stickiness in the indicator needle. As a result, his climb angle was low and he did not get high enough to conduct the onboard experiment. The flight back to Edwards was uneventful; he had finally gotten a nice peaceful flight. Maybe Adams' luck was improving.

Adams' fifth flight was in the number one aircraft again. It was his third altitude build-up flight. The planned maximum altitude was 220,000 feet, only 46,000 feet shy of the altitude needed to get his astronaut wings. The flight was flown as planned.

During his sixth flight, in the number three aircraft, the rocket engine failed to ignite on launch from the B-52. Adams managed to get the engine lit on his second try, sixteen seconds after launch, and finally got back on the planned profile. Due to a problem with a computer, the boost guidance experiment was canceled, but he obtained good data on three other experiments.

While Adams was flying the X-15, I completed three stability and control flight test programs in the F-4 Phantom. TAC headquarters requested that I evaluate the Phantom during weapons delivery passes with a variety of armament. During certain flight conditions, I found that the F-4 had some extremely dangerous flying qualities, which could cause loss of control and a possible crash. My flight test engineer and I came up with some simple modifications to the Phantom. These changes should lessen the risk. I also found methods of flying the aircraft that gave the pilot a better way of determining how close he was to the danger zone. TAC headquarters wanted me to give presentations on the results of my tests to other Phantom squadrons all over the United States. So every Friday afternoon, I gave a safety briefing to a different F-4 squadron just before happy hour at the Officers Club.

Interest in my F-4 flight test results also came from the headquarters of the United States Air Forces in Europe (USAFE). A request came for me to brief the European headquarters staff and all the F-4 squadrons in England, Germany, and the gunnery range at Wheelus Air Base, Libya. The Flight Test Center prepared rush orders for me to fly in a C-141 Starlifter to arrive in Europe as soon as possible. Before leaving for Europe, I asked Mike Adams to autograph a photograph of himself standing next to the X-15 and he agreed to send me one.

Adams' seventh X-15 flight took place in the number three aircraft on November 15, 1967, while I was in Europe. After drop from the B-52, the X-15 rocketed into and out of the atmosphere. He reached a peak altitude of 266,400 feet, qualifying as the twenty-seventh U.S. astronaut. This was the 191st flight in the X-15 program, sixty-fifth flight of the X-15 number three, seventh flight for Adams, and his third in this aircraft. He trained for this flight for twenty-three hours in the X-15 simulator and practiced portions of the profile in an F-104.

The rocket-powered climb portion of his flight was normal except for an aircraft electrical system disturbance attributed to arcing of one of the onboard experiments. The disturbance affected the quality of the information radioed to the ground, the altitude and velocity computer and cockpit displays of the inertial guidance system, and the automatic operations of the primary control system. Although this did create a distraction to Adams and degraded the automatic feature of the reaction control system, he possessed adequate flight instruments to complete the flight.

As the X-15 approached peak altitude, the aircraft began a slow change in heading to the right at a rate of one-half degree per second. Because this portion of the flight was outside most of the atmosphere and was, in effect, coasting along a predetermined flight profile, only the heading (direction of the plane) and not the flight path changed.

The automatic part of the reaction control system operated normally for short periods, and near peak altitude, 266,400 feet (50.4 miles), the X-15's heading was 15 degrees right of flight path.

Possibly suffering from disorientation and vertigo, Mike Adams apparently mistook a roll indicator for a sideslip (heading) indicator and used manual reaction control to drive the aircraft farther to the right in heading. The indicator could be selected by the pilot to indicate either sideslip or roll. Cockpit films show the needle was indicating roll as was planned. Three other instruments in the cockpit were all correctly indicating the error in heading and sideslip.

About thirty seconds after descending from peak altitude and as aerodynamic forces were becoming more intense due to denser air, the X-15 heading error increased to 90 degrees right of flight path. The plane entered a spin at an altitude of approximately 230,000 feet and a speed of Mach 5, more than 3,000 mph.

The aircraft continued to spin some forty-three seconds down to an altitude of about 120,000 feet where some combination of Adams' action, the X-15's inherent stability, and the stability augmentation portion of the Honeywell MH-96 control system caused the aircraft to recover from the spin. However, immediately following the spin recovery, the X-15 developed a pitch oscillation (nose up-and-down motion). The automatic control system became saturated, causing the pitch damper (an electronic device that put in flight control movement in addition to the pilot input for additional aircraft stability) gain to remain at maximum setting. Because of this high-gain setting, the automatic control system caused the oscillation to become self-sustaining and to increase in severity.

Once the control system oscillation began, the only means of stopping the oscillation and recovering the aircraft was to reduce the amount of pitch gain to minimum or to turn the pitch damper off. Neither of these actions was taken prior to the loss of telemetry. It is doubtful that Adams could have recognized the problem and activated the proper switches under the heavy stresses of the flight and with the violent aircraft motions that he was experiencing. Even Pete Knight, who was the NASA flight controller monitoring the flight, misread the telemetry strips and kept telling Adams to pull some Gs.

At this time the aircraft was descending at approximately 160,000 feet per minute (2,600 feet per second), and the dynamic pressure was increasing at nearly 100 pounds per square foot per second. There was a corresponding increase in acceleration forces, with both positive and negative forces in excess of +15G, which exceeded the structural limitations of the X-15. The airplane broke into many pieces while still at high altitude in excess of 60,000 feet. Adams, probably incapacitated by the high-G forces, did not escape from the X-15 cockpit and was killed on ground impact. His nine lives were up.

I was in Europe giving briefings on my tests in the F-4 Phantom when Adams was killed. During a refueling stop in Germany, I obtained a copy of *Stars and Stripes*, a daily newspaper printed overseas with news for military personnel. On the cover of the newspaper was the headline, "Mike Adams Killed in the X-15." This was the first knowledge I had of his accident. I called my boss at Edwards to learn the details and found out I had received orders to fly combat in Vietnam. I personally felt a great loss in Adams' death and regretted that I was five thousand miles away from Edwards.

Returning home from Europe, I went to Test Operations to read correspondence that had arrived while I was gone. On my desk was a huge pile of flight manual revisions, military Teletype messages, and a brown manila envelope. In the envelope was an autographed photograph of Mike Adams standing next to the rocket-powered X-15.

My new orders called for me to fly the Douglas A-1 Skyraider in the 602nd Fighter Squadron (C)—C for Commando—from Udorn Royal Thai Air Force Base, Thailand. The squadron's mission was to lead Sikorsky HH-3 Jolly Green helicopters in the rescue of Air Force and Navy pilots shot down over the Ho Chi Minh Trail in North Vietnam and northern Laos. Before leaving Edwards I completed my test reports and prepared to move my family to Hurlburt Air Force Base in Fort Walton Beach, Florida. There I would learn to fly and fight in the Skyraider.

Before leaving Edwards I was scheduled to attend the Air Force Escape & Evasion (E&E) School at Fairchild Air Force Base near Spokane, Washington. Capt. Ronald V. "Ron" Franzen, one of my classmates at the test pilot school, had also been given orders to Vietnam. We departed Edwards in early December for a week in the E&E School. During the course of training for aircrew, we were subjected to conditions we could expect if we were shot down and captured. Our captors seemed very authentic, but they were actually U.S. servicemen dressed in Russian uniforms. Franzen and I were given a small taste of the horrors that could be inflicted on us if we were unfortunate enough to be imprisoned by the North Vietnamese or the Laotian Pathet Lao.

When I returned to Edwards after attending E&E School, I found that tragedy had again struck the Flight Test Center. On the afternoon of December 8, 1967, Maj. Robert H. "Bob" Lawrence Jr. was killed. Lawrence, thirty-one, was the first African American astronaut-designee. He had graduated six months earlier from the test pilot school and been selected as a MOL astronaut. Lawrence was born and raised in Chicago, where he excelled in school. At Bradley University he signed up for ROTC, became the corps commander, and graduated in 1956 with a degree in chemistry and an Air Force second lieutenant's commission.

After completing flight training at Malden Air Base in Missouri, Lawrence was assigned as an instructor pilot for the German air force, flying T-33 trainers at Fürstenfeldbruck Air Base near Munich. He then earned a PhD in physical chemistry at Ohio State University, completing his dissertation in 1965. After graduation Lawrence remained an active jet pilot while performing research at Kirtland Air Force Base and eventually accumulated more than twenty-five hundred hours of flight time. Much of his flying was done at Edwards, where he graduated from the test pilot school in mid-1967. On June 10 Lawrence was officially selected a crew member candidate for the MOL project, one of the pilots selected in the third group. He and his MOL teammates remained at Edwards performing research and training tasks.

Throughout the 1960s both NASA and the Air Force were experimenting with small-winged aircraft such as the X-15 rocketship and even small-winged flying craft lifting bodies that were seen as prototypes and test beds for future manned space vehicles. At the time planners had sought ways to simulate the in-flight mechanics of space and reentry vehicles. The skills and experience required to master those mechanics were considered necessary for test pilots and astronaut trainees. Early on, while training pilots to fly the X-15, instructors noted that the F-104 Starfighter in an altered configuration (i.e., with speed brake extended and power to idle) closely resembled the X-15 in unpowered flight. As a student I climbed to 25,000 feet directly over Runway 04 at Edwards, pulled the power to idle, extended the speed brakes, and started a 360-degree turning descent at 350 knots. The plan was to arrive 500 feet above the desert floor aimed at the runway, extend the landing gear, and touch down on a white stripe painted across the 10,000-foot remaining point of the 15,000-foot runway. With a little practice I could land within one hundred feet of the line every time.

A much more difficult approach was designed to simulate a space vehicle with a lift-to-drag ratio of 2.5. In this case the F-104 was again flown to 25,000 feet. But instead of positioning the F-104 directly over Runway 04, the plane was positioned about ten miles from the end, just west of Pancho Barnes' former Happy Bottom Flying Club. For this "dirty" approach the power was placed in idle, speed brakes were out, with landing gear and takeoff flaps extended. A 30-degree dive was established at 295 knots, the maximum speed allowed with the landing gear down. About 1,500 feet above the desert, the pilot started a flare and the speed rapidly bled off. Again the plan was to touch down on the white line. This approach was extremely difficult to perform since the initial conditions determined the touchdown point, not pilot technique. If the approach was started beyond ten miles, the aircraft did not have enough energy to make it to the runway. If the approach was started too close, the dive angle was excessive and the pilot might not be able to pull out of the dive. Instructors told us if the approach looked bad early on, just add power and go around.

Maj. Harvey J. Royer, USAF, and Maj. Robert Lawrence left the test pilot school briefing office and walked out to F-104D serial number 57-1357. Major Royer, the school's chief of operations, was giving Lawrence an upgrade to instructor pilot. Their mission was a standard series of approaches practiced by every pilot in the school. The plan called for them to make two simulated X-15 approaches until fuel depleted to 3,500 pounds, two "clean" (i.e., gear-up) low lift/drag approaches until fuel burned to 2,500 pounds or less, and two "dirty" approaches.

As they went through the drill that afternoon, things went horribly wrong for Lawrence and Royer. They were making a "dirty" approach and for some reason lost control of the Starfighter. No one will ever know for sure why, but the F-104 smacked into the runway left of centerline, 2,200 feet down Runway 04 (2,800 feet short of the white line), the Starfighter's underbelly blossoming fire. The main landing gear collapsed on first contact and the canopy shattered. The Starfighter's fuselage dragged along the runway for over 200 feet, then took to the air again, sailing madly down the runway for another 1,800 feet. Royer ejected and Lawrence also punched out. At the 11,000-foot remaining marker, the ship veered left, and, 235 feet later, the twisted F-104 wreckage left the runway and skidded to a stop in the sand. Royer was seriously hurt. Lawrence's shattered body landed seventy-five yards from the wreck. He was still strapped in his ejection seat, the parachute unopened. His chest was crushed and his heart lacerated.

Major Lawrence died instantly, less than six months after being congratulated by the Air Force Chief of Staff Gen. J. P. McConnell for his selection to the MOL program. In a letter to Lawrence, General McConnell wrote, "Ahead of you lies adventure, challenge and an opportunity to serve the Air Force and the nation such as few men have had. Your tasks and responsibilities will be extremely demanding. I am certain you will measure up to them." Maj. Robert Henry Lawrence Jr. gave full measure; he was the ninth U.S. astronaut to die.[3]

Ebony staff writer David Flores wrote, "Robert Lawrence was a hero in the true American tradition who in so doing proved, finally, what black people in this country have long known—that excellence has no color. For young black boys . . . should they seek to be heroes in the traditional manner or of another kind, it is the legacy one courageous black man left them."[4]

Colonel Chuck Yeager had ejected from an NF-104 Starfighter and been injured just two weeks before I arrived at Edwards in late December 1963 to attend the test pilot school. He was trying to set a world altitude record when the mishap occurred. Now Maj. Robert Lawrence had just been killed in another F-104 Starfighter while practicing approaches to be used to bring winged vehicles back from space. As I prepared to depart Edwards for my tour of combat in Vietnam, it seemed that events had gone full circle in the four years I had spent in flight test. I was saddened by the loss of so many test pilots, but I was proud of what I had accomplished during my test flights over the Mojave Desert.

Epilogue

I departed Edwards in December 1967 and spent a year flying combat in Vietnam. I was a Douglas A-1 Skyraider ("Sandy") rescue pilot in the 602nd Fighter Squadron (C) (C for Commando), flying from Udorn and Nakhon Phanom Royal Thai Air Force bases in Thailand. It was a very difficult year for me because twelve of my squadron mates were killed, two were burned so badly they were sent home, and another was injured so severely on ejection that he was medically retired. My squadron also lost twenty-six aircraft during that horrific year in our country's history. I completed 188 combat missions, logged over 600 combat hours, and was awarded the Distinguished Flying Cross with two oak leaf clusters and the Air Medal with eight oak leaf clusters. A book about my experiences, *Cheating Death: Combat Air Rescues in Vietnam and Laos*, was published by the Smithsonian Institution Press in 2003.

While I was flying combat in Vietnam in 1968 and 1969, I followed the flight test activities at Edwards. Many flight test programs were either completed or canceled, and the aircraft used to set world speed and altitude records were sent to museums or placed in salvage. Bill Anders, my friend from the 84th Fighter Interceptor Squadron at Hamilton Air Force Base, flew to the moon on *Apollo 8* over Christmas in 1968, setting speed and altitude records far surpassing anything remotely possible at Edwards. The golden age of flight testing over the Mojave Desert had come to an end.

After about ten years of record-setting flights, the X-15 test missions were completed. My NASA test pilot friend William "Bill" Dana, thirty-eight, from Pasadena, California, flew the last X-15 flight on October 24, 1968. The number one rocketship was transported to the National Air and Space Museum in Washington, D.C., and the number two aircraft was taken to the National Museum of the United States Air Force at Wright-Patterson Air Force Base in Dayton, Ohio. Mike Adams had been killed in the crash of the number three aircraft in 1967. Only three X-15s were ever built.

The one remaining XB-70 Valkyrie, the number one aircraft, was also retired. Fitzhugh "Fitz" Fulton, forty-three, from Blakely, Georgia, and copilot Emil "Ted" Sturmthal, both test pilot friends of mine, ferried Cecil the Seasick Sea Serpent to its final resting place at the National Museum of the United States Air Force on February 4, 1969. Sturmthal is credited with saying, "I'd do anything to keep the Valkyrie flying—except pay for it myself."

Secretary of Defense McNamara canceled the YF-12 Blackbird program, as he did the Dyna-Soar program five years earlier. The three YF-12s, the only ones ever built, remained at Edwards and were flown occasionally. Lt. Col. Ronald "Jack" Layton, forty-four, from the White Mountains region of Arizona, was a pilot in the 84th Fighter Interceptor Squadron at Hamilton Air Force Base when I was stationed there in the early 1960s, and later he was a CIA pilot flying the A-12. On June 24, 1971, he ejected from the number three YF-12 when one engine caught fire. Layton and his weapon systems officer survived with only minor injuries. The number two YF-12 aircraft was seriously damaged during a landing at Edwards and was placed in storage in Palmdale, California. The rear half of the plane was later used to build the SR-71C trainer. The number one YF-12 was flown to the National Museum of the United States Air Force for permanent display on November 7, 1979.

Flight testing on the SR-71 was completed at Edwards and all the Blackbirds transferred to the 9th Strategic Reconnaissance Wing at Beale Air Force Base, California. The triple-sonic SR-71 performed reconnaissance missions for about twenty years, but the Blackbird has now been retired for over ten years. Lockheed built thirty-two of the record-setting Blackbirds. Twenty are on public display at museums around the country.

Six F-111A aircraft were sent to Vietnam in March 1968 in a combat evaluation called Combat Lancer. Secretary McNamara wanted to demonstrate that his decision to select an aircraft to be used by both the Air Force and Navy was superior by deploying the aircraft in combat as soon as possible. Unfortunately, the first aircraft crashed only eleven days after it arrived in Vietnam. Another crashed two days after the first one. A third vanished a short time later. To lose half of the F-111A force in a month was a disaster. It was not possible to get even a single clue about the first two losses, as the crews maintained radio silence on single-ship night combat missions. No wreckage from either aircraft was ever found. Neither aircraft appeared to have been shot down in the target area, a belief strengthened by the report of the third crew, which ejected. The F-111 was a dismal failure in combat.

The number eight F-111A, serial number 63-9773, and the first of the two F-111A aircraft I flew at Edwards, is now on static display at the air park on Sheppard Air Force Base near Wichita Falls, Texas. The other F-111A

I flew, serial number 63-9774, crashed in January 1967, killing Maj. Herb Brightwell. It is my belief that the F-111A was a terrible failure in design, in flight test, and in combat operation. The plane also caused the death of 3 of my test pilot friends and another 103 Air Force pilots.

The F-4 Phantom was already being used in combat in Vietnam while I was testing it at Edwards. Some of my recommendations were incorporated in the fighter, but many were not. The F-4 was the workhorse in Vietnam but suffered a high loss rate. The Air Force lost 444 Phantoms in that conflict. I believe at least half of the losses were due to pilot loss of control while maneuvering—a needless loss after all the testing I did at Edwards and warnings I gave to senior Air Force management.

As the years rolled by I kept track of some of the pilots I knew in flight school, in combat, and at Edwards through membership in the Society of Experimental Test Pilots, social gatherings, and occasionally newspaper articles. I was shocked to find information about some of my friends.

Lenard Kresheck, my Air Force flight instructor at Webb Air Force Base, Texas, in 1958 and 1959, left the military service and joined Trans World Airlines (TWA). He had initially started with TWA as a flight engineer on the Boeing 727, then upgraded to first officer or copilot. In late 1974 I read an article in the *Washington Post:*

> TWA Jetliner Crash
> A TWA jetliner, flying here from Indianapolis and Columbus, Ohio, in fierce winds and heavy rain, crashed and burned in the Blue Ridge Mountains 47 miles west of Washington on its approach to Dulles Airport yesterday killing all 92 persons aboard.
>
> Flight 514, a three-engine-jet Boeing stretch 727 with 85 passengers and a crew of seven, slammed into the western slope of the mountains 23 miles west of Dulles and four miles south of Rte. 7 in Loudon County at about 11:10 a.m.
>
> The aircraft, which had been diverted from National Airport because of weather, sheared off treetops, struck a rocky outcrop, broke up and caught fire, scattering charred bodies and parts of bodies over an area about the size of two football fields.
>
> The flight crew consisted of Richard Brock, Captain, Lenard Kresheck, First Officer and Tom Safranek, Flight Engineer.[1]

On December 1, 1974, the crew of TWA Flight 514 took off from Columbus, Ohio, at 10:24 AM bound for Washington National Airport. Cleveland Air

Traffic Control gave Flight 514 the message that no landings were allowed at National due to strong surface winds. Aircraft were being diverted to Dulles International. The plane reached its cruising altitude of 29,000 feet, and flight engineer Safranek talked to the TWA dispatcher in New York on the company radio. The dispatcher told them to plan on terminating at Dulles. First officer Lenard Kresheck, in the copilot seat, complained about the rough ride the automatic pilot was giving them as he pulled out the approach plate for Dulles. He studied it briefly and asked Captain Brock for the wind forecast. Kresheck disconnected the autopilot and started to fly the aircraft himself. Brock asked whether anyone had told the passengers about the rerouting.

Ground control had directed Captain Brock to cross the 300-degree radial twenty-five nautical miles northwest of Dulles Airport at 8,000 feet. Brock repeated the instructions and Kresheck acknowledged them. Pulling out the Dulles approach plate, the captain studied the letdown information.

Kresheck concentrated on keeping the jet on course as Safranek read out the preliminary landing checklist. The two senior men answered, flipping switches and turning knobs. Kresheck made a small course correction, lining up the 727's nose on the VOR (visual omni range) radio signal, which extended directly down Runway 12 at Dulles.

Brock was concerned with the weather and with landing on an unfamiliar strip. The approach plate, a rectangular sheet of thick white paper crisscrossed with lines and numbers indicating air routes and minimum altitudes for those routes within a twenty-five-mile radius of the airfield, was "the bible." A pilot must fly the approach exactly as depicted, with no deviations.

The plane descended 11,000 feet toward more turbulent air. The crew listened as Dulles approach control gave them a rundown on weather conditions on the ground. Moments later, Dulles cleared TWA 514 for a VOR DME (distance measuring equipment) approach to Runway 12.

It was 11:04 AM on December 1, 1974, and Richard Brock and Lenard Kresheck had just made the biggest mistake of their flying careers—one that caused them to perish. They started to descend to the final approach altitude miles before their clearance limit.

The four-and-a-half-year-old plane shuddered through the cloud layer, with snow and hail alternately streaking the cockpit windows. The seat-belt sign was on—even the flight attendants were strapped in. Brock spotted the gray waters of the Shenandoah below and slightly ahead. They were over a river.

The plane lurched, and Kresheck fought for control. The radio altimeter horn sounded, then stopped. It was preset to go off 500 feet above the ground. The broad, level plain leading to Runway 12 would have long since come into view had the plane not been flying in clouds.

The foothills of the Blue Ridge triggered the radio horn. At 198 knots the screaming jetliner, left wing slightly dipped, hit the first black oak poking seventy feet up through the mist of Weather Mountain.

Within the next fraction of a second, like a giant lawn mower out of the sky, the jetliner cut a 380-foot swath through pine, spruce, and oak, ripping and tearing its way lower and lower toward Route 601.

The men in the cockpit were already dead—speared by tree limbs shattering the windscreen. The jet crossed the road at fifteen feet, sinking. Nose first, it bludgeoned into the basalt ledge on the far side of Route 601 and quite simply ceased to exist. A bloody rain of limbs and debris first rose and then fell through the bare branches. Fire spread on the rock. The fog and silence then rolled back in.[2]

It was very painful for me to research the flight data on TWA 514. Even though I had known Lenard Kresheck for only six months, I admired and respected him immensely. He taught me how to fly the T-33 and stood next to me when Jan pinned the Air Force silver wings on my uniform. Kresheck recommended to the Air Force pilot selection board that my name be added to a list for jet fighter training. He was a vital part of my dream coming true and another step in chasing the whistle.

NASA selected Capt. **Bill Anders** as an astronaut in 1963. Because of his master's degree in nuclear engineering, his engineering responsibilities at NASA were dosimetry, radiation effects, and environmental controls. Anders was selected as backup pilot for the *Gemini 11* mission and backup command module pilot for *Apollo 11.*

Anders was the prime lunar module pilot for *Apollo 8*, the first lunar orbit mission, in December 1968. Frank Borman, Jim Lovell, and Anders were the first humans to leave the gravitational pull of the earth on their five-day trip to the moon. While in moon orbit, Anders took photos of potential landing spots. He also took a photograph known as "Earthrise" (the earth rising above the moon), which has become one of the most celebrated pictures of the space program, and which was later used on a U.S. postage stamp. The famed nature photographer Galen Rowell has called this image "the most influential environmental photograph ever taken."

From June 1969 to 1973 Anders served as executive secretary for the National Aeronautics and Space Council, which was responsible to the president, vice president, and cabinet-level members of the council for developing policy options concerning research, development, operations, and planning of aeronautical and space systems.

Anders was then appointed to the five-member Atomic Energy Commission, where he was lead commissioner for all nuclear and non-nuclear power research and development. He was also named as U.S. chairman of the joint US/USSR technology exchange program for nuclear fission and fusion power.

Following the reorganization of national nuclear regulatory and developmental activities in 1975, Anders was named by President Gerald Ford to become the first chairman of the newly established Nuclear Regulatory Commission (NRC) responsible for nuclear safety and environmental compatibility. At the completion of his term as NRC chairman, Anders was appointed U.S. ambassador to Norway and held that position until 1977. He continued to fly the T-38 Talon while he was ambassador.

Anders left the federal government after twenty-six years of service and joined General Electric as vice president and general manager of the Nuclear Products Division in San Jose, California. In January 1980 he was appointed general manager of the GE Aircraft Equipment Division with headquarters in Utica, New York. With more than eighty-five hundred employees in five plants in the northeast, Aircraft Equipment Division products included aircraft flight and weapon control systems, cockpit instruments, aircraft electrical generating systems, airborne radars and data processing systems, electronic countermeasures, space command systems, and aircraft/surface multibarrel armament systems. In 1984 he left GE to join Textron as executive vice president, aerospace, moving to senior executive vice president, operations, in 1986. He was also a consultant to the Office of Science and Technology Policy and was a member of the Defense Science Board and the NASA Advisory Council.

Anders became vice chairman of General Dynamics Corporation in 1990 and then chairman and chief executive officer in January 1991. In 1993 he retired from the company but remained chairman of the board until May 1994, when he fully retired.

Bill Anders, a retired major general in the Air Force Reserve, is now the chief pilot, founder, and president of the Heritage Flight Museum located at Bellingham International Airport in the state of Washington. He is one of the very few pilots that have been approved by both the Air Force and Navy to fly in formation with military demonstration team aircraft during air shows. With over eight thousand hours of flying time, he flies his P-51D Mustang *Val-halla* with F-16, F-15, and A-10 fighter aircraft in the Air Force Heritage Flight aerial demonstration program. He also flies his F8F-2 Bearcat *Wampus Cat* with F-18 fighter/attack aircraft in the Navy Tailhook Legacy Flight aerial demonstration program. Since the inception of his museum, Bill Anders has returned to what he likes best—flying.[3]

After graduation from test pilot school in December 1964, my classmate **Jim Hurt** was assigned to the Air Force Special Weapons Center at Kirtland Air Force Base, New Mexico. Hurt moved with his wife and two daughters to Albuquerque, but his marriage ended shortly afterward. Since he had been a T-38 instructor pilot in Air Training Command prior to the school, he was pleased to get the opportunity to fly fighter-type aircraft. Hurt spent the next three years at Kirtland flying the F-100, F-104, and F-105 aircraft, testing delivery techniques to drop nuclear weapon shapes.

As the war in Vietnam escalated, Hurt got orders to fly combat. He was sent to George Air Force Base, California, to check out in the F-4 Phantom. Six months later he was transferred to the 13th Tactical Fighter Squadron at Udorn Royal Air Force Base in Thailand. When I first arrived at Udorn to start my year of combat flying the Douglas A-1 Skyraider, I met Hurt in the officers club. He was excited to qualify to fly fighter aircraft and had become an aircraft commander in the F-4. While in Southeast Asia Hurt was awarded the Air Medal with eight oak leaf clusters, the Distinguished Flying Cross, and the Bronze Star. He completed his tour as a duty officer at 7th Air Force headquarters at Tan Son Nhut Air Base, South Vietnam.

Jim Hurt's life took a very positive turn when he returned home after service in Vietnam. He married Ann Wheeler, was promoted to major, and was assigned to the Fighter Branch of Test Operations at Edwards. It was a dream come true for him to finally qualify to test fighter aircraft at the Flight Test Center and to have a stable marriage. A short time later his wife was expecting their first child.

Soon Hurt checked out in the General Dynamics F-111E, the advanced version of the F-111A that I had flown at Edwards in 1966 and 1967. While Hurt was testing aircraft at Edwards, four models of the F-111 were being evaluated: the F-111A, D, and E; and the FB-111A. The F-111A was the basic airplane, the all-weather tactical fighter-bomber. The D model differed from the A model by having improved Mark II avionics. The E model had an extensively modified engine air inlet and better ground-attack penetration aids. The FB-111A was the Strategic Air Command bomber version, with the F-111A's fuselage, extended wingtips, a more powerful engine, and the Mark IIB improved avionics package.

Jim Hurt flew seven test flights totaling twenty-eight flight hours in a YF-111D aircraft equipped with E avionics. The Mark II avionics system had microelectronic circuitry for improved reliability and power and reduced weight. Hurt flew three different navigation routes to evaluate the avionics over a wide range of terrain features including desert, mountains, and land-water contrast.

Five years after I had been involved in testing the F-111A at Edwards, the aircraft continued to be controversial, and its highly publicized crashes directed the attention of the Air Force and news agencies to its testing. The Air Force Joint Test Force had expanded to two hundred people and was directed by a full colonel. The F-111 was an aircraft that started out against the odds. No previous USAF production aircraft had embodied so many unknowns at one time—new variable-sweep wings combined with new and untried power plants; hitherto unapproached advances in afterburning; radically new avionics and fire-control systems; and completely new missile systems—and the whole wrapped in the concept of commonality. The criticism was well known: the concepts of variable geometry and commonality were invalid; the choice of contractor was the result of political considerations; the cost escalation of the program was excessive; the F-111 suffered from excessive flight restrictions; it was seriously deficient in safety and performance; and the aircraft turned out to be a failure operationally. Nevertheless, Jim Hurt was optimistic about the F-111.

In March 1971 Hurt was asked to talk about the aircraft at the Antelope Valley Board of Trade breakfast meeting. The local *Ledger-Press* reported on what Hurt said:

> Publicity on the aircraft has given it a bad reputation. As far as safety records, the F-111 outshines other century series jets at the 15,000 hour mark and is continuing to hold its own at the 45,000 hour mark. Nine F-111s were lost in the first 15,000 hours of flight as compared to 14 losses in the F-105 program, 15 in the F-102 program, 21 in the F-104, and 28 in the F-100. At the 45,000 hour mark, only 12 of the F-111 aircraft had been lost.
>
> The only negative point is that the side-by-side seating arrangement of the pilot and fire control officer made the classic dogfight situation a poor one for the F-111 because of decreased visibility from the pilot's seat.[4]

About a month later, on April 23, 1971, Jim Hurt and his copilot, Maj. Robert J. Furman, took off from Edwards in an F-111E, serial number 67-117. The flight's purpose was to evaluate operation of the gun during flight from 300 to 400 knots airspeed, at 20,000 feet, and during turning maneuvers up to 3 Gs. In the past, frequent engine compressor stalls had occurred during gunfire. Also while performing accelerated flight, the F-111E had occasionally stalled and gone into a spin. Hurt and Furman were flying an extremely dangerous mission.

After takeoff, Hurt flew northeast of Edwards to the Leach Lake Gunnery Range with Major Myers, the pilot of an F-104D Starfighter, acting as a photo chase. The first firing pass was made eighteen minutes after takeoff, at which time the chase pilot reported he could not hold position. The F-104 was quite limited in its capability to pull Gs at low speed. Hurt commented to Major Furman that his airspeed had decreased to 300 knots during the maneuver; he had experienced an engine stall and would repeat the test.

Hurt started his second gun-firing pass in a right turn. Suddenly Major Myers saw the F-111 in a left bank and realized the aircraft was out of control.

"Are you all right?" Myers asked.

"I don't know," Hurt calmly replied.

"Watch your altitude," called Myers as the F-111E rapidly descended in altitude.

"Watch your altitude!" repeated Myers, and three seconds later he said, "Get out!"

The F-111E was equipped with a crew module that allowed Hurt and Furman to eject together from the stricken plane. Major Myers saw the crew module fire off the nose of the aircraft. He waited for a huge parachute to deploy and safely lower the two pilots to the desert floor. Unfortunately, a plate covering the parachute container failed to blow off the capsule. As a result the crew module dived toward the desert and crashed about one-quarter mile west of the F-111E impact point. Maj. Jim Hurt and Maj. Robert Furman were instantly killed. Hurt's son, Jeffrey, was born six months later.[5]

After graduating from test pilot school, my classmate **Fred Haise** was one of nineteen NASA astronauts selected in April 1966. He served as backup lunar module pilot for *Apollo 8* (behind Bill Anders). Haise also was the backup lunar module pilot for Buzz Aldrin on the *Apollo 11* first flight to the moon mission.

Haise's first and only space flight was aboard the ill-fated *Apollo 13* spacecraft on which he served as the lunar module pilot. He was scheduled to be the sixth man to walk on the moon. The mission, launched April 11, 1970, and programmed for ten days, was committed to the first landing in the hilly, upland Fra Mauro region of the moon. However, fifty-five hours into the flight, a failure of the service module cryogenic oxygen system occurred, which led to an explosion and an emergency departure from the original flight plan. Fred Haise and fellow crewmen Jim Lovell (spacecraft commander) and John L. Swigert (command module pilot), working closely with Houston ground controllers, converted their lunar module *Aquarius* into an effective lifeboat. Unfortunately, about two days into the flight Haise felt a fever coming on with lightheadedness, clammy skin, and tingly nerve endings. To conserve their remaining power, the space capsule's heating was turned down, causing the temperature to drop to the low 40s. For the remaining flight, Haise was miserable and shaking from the cold. The crew's emergency activation and operation of lunar module systems conserved both electrical power and water sufficiently to ensure their safety and survival while in space for eighty-six hours. After the first deep-space abort in history, the crew returned to earth safely on April 16, 1970.

Haise next served as backup spacecraft commander for the *Apollo 16* mission, and he would have commanded the *Apollo 19* lunar landing had that mission not been canceled.

In 1973, while flying a replica of a World War II Japanese aircraft for the Confederate Air Force, Haise crashed and was badly burned. He recovered from his injuries and continued working for NASA on the Space Shuttle program. From April 1973 to January 1976 he was technical assistant to the manager of the Space Shuttle Orbiter Project.

Haise was the commander of one of the two two-man crews that piloted the orbiter *Enterprise* during the approach-and-landing test (ALT) flights from June to October 1977. This series of critical orbiter flight tests initially involved Boeing 747/orbiter captive-active flights, followed by air-launched unpowered glide, approach, and landing tests (free flights). There were three tests with the orbiter *Enterprise* carried atop the Boeing 747 carrier aircraft, allowing in-flight, low-altitude, and low-speed test and checkout of flight control systems and orbiter controls. Five free flights were flown, which permitted extensive evaluations of the orbiter's subsonic flying qualities, performance characteristics during separation, up-and-away flight, flare, landing, and rollout. Haise commanded free flights 1, 3, and 5 and was the backup commander for free flights 2 and 4.

In March 1978 Haise was assigned to command the third planned shuttle orbital flight test, which was scheduled to rendezvous with the Skylab station then in orbit. But technical problems delayed the first shuttle flight past Skylab's reentry and destruction on July 11, 1979.

On his lone space flight, Fred Haise logged 142 hours and 55 minutes in space. He was disappointed that he never got a chance to return to the moon. A combination of bad luck and ill timing prevented him from getting another trip, the greatest scientific epic journey of our century. He told me he felt cheated after all the years he had trained for a space mission to never get the opportunity to go back to the moon. Haise had come close to realizing his lifelong dream of walking on the moon—and of comparing his zero-G experience in the Vomit Comet to the moon's 1/6th of earth's gravity.[6]

Capt. **Jim Taylor,** the unofficial winner of the Fleet enema contest, was in the class before me at the Test Pilot School. He graduated six months earlier and was assigned to the Air Force Special Weapons Test Center at Kirtland Air Force Base. For a year he flew as a test pilot in the F-106 Delta Dart, developing the Hughes Aircraft Company's MA-1 fire-control system.

When the start of the Air Force Manned Orbiting Laboratory was announced in the fall of 1965, Taylor was selected as one of the first eight MOL astronauts and promoted to major. He remained with the program until it was canceled in 1969. Likewise, the first Air Force manned space program, called Dyna-Soar, had been canceled in 1963. The Air Force now had two manned space programs canceled by the Department of Defense. They never got the opportunity to try again—the road to space for an Air Force pilot started in Houston.

Meanwhile, Taylor was promoted to lieutenant colonel and offered the chance to become deputy commandant of the Test Pilot School. He accepted the assignment, and he and his wife Jacquelyn and their three children returned to Edwards and once again moved into base housing.

On September 4, 1970, Taylor was scheduled to instruct Captain Pierre J. du Bucq, a French air force exchange test pilot trainee, in the T-38. Captain du Bucq had flown his first flight in the Talon with Taylor a few days earlier.

Taylor and du Bucq flew the same checkout procedure I had flown in the T-38 six years earlier. They made a maximum-power takeoff, climbed to 35,000 feet, performed acrobatics, practiced single-engine flight at 20,000 feet, and went vertical.

Using the call sign School 8-02, they notified Edwards approach control that they were switching to Palmdale tower. On contact with Palmdale tower, they requested touch-and-go landings and were cleared to enter on initial for the 10,000-foot Runway 25 for traffic sequencing with other aircraft in the pattern. At this time, Palmdale was reporting calm winds and an altimeter setting of 29.78. Sky conditions were clear with thirty miles visibility.

School 8-02 completed one normal overhead pattern and a touch-and-go landing to Runway 25 and then was cleared for closed traffic. The Air Force closed-traffic procedure consisted of accelerating to takeoff speed after a touch-and-go landing, retracting gear and flaps, then flying to the far end of the runway and pulling up into a downwind leg at pattern altitude.

Following School 8-02's gear check on base turn, Palmdale tower cautioned them of wake turbulence from a C-141 Starlifter that was conducting touch-and-gos on the Palmdale 10,000-foot Runway 22. The first 4,000 feet of Runway 22 at Palmdale intersected with and was below the approach end of Runway 25 on which School 8-02 was flying its practice landings.

When Taylor and du Bucq were on final approach and only a few feet in the air, they encountered severe wake turbulence from the huge C-141 transport. Their Talon rolled over and crashed about one thousand feet short of the end of Runway 25. Firefighting and rescue personnel responded rapidly, within one to two minutes, and quickly extinguished a small fire in the T-38's tail section with firefighting foam. Jim Taylor and Pierre du Bucq were, however, fatally injured.[7]

Maj. **Joseph "Jovial Joe" Engle,** the pilot who pulled so many Gs in a T-38 that I became unconscious, was selected as a NASA astronaut in 1966. He moved to Houston and looked forward to taking the trip to the moon for which he had always longed. NASA selected him as the backup lunar module pilot for the *Apollo 14* mission, the fourth flight to the moon. Unfortunately for Engle, the prime crew made the flight and successfully landed on the moon.

Later he was selected as the prime lunar module pilot for *Apollo 16.* Due to budget constraints, *Apollo 17* was canceled and geologist Harrison Schmidt, the lunar module pilot for that mission, was moved forward one flight and replaced Engle on *Apollo 16.* As a result, Engle never made a flight to the moon.

Engle was commander of one of the two crews that flew the space shuttle *Enterprise* approach-and-landing test flights from June through October 1977. The *Enterprise* was carried to 25,000 feet atop a Boeing 747 carrier aircraft and then released for its two-minute glide flight to landing. In this series of flight tests, Engle evaluated the orbiter handling qualities and landing characteristics and obtained stability and control and performance data in the subsonic flight envelope for the space shuttle. Lt. Dick Truly, USN, a classmate of mine in the test pilot school, flew as Engle's copilot.

Joe Engle was also commander of the second orbital test flight of the Space Shuttle *Columbia*, which launched from Kennedy Space Center, Florida, on November 12, 1981. His copilot for this flight, which was called STS-2, was again Dick Truly. Despite a mission shortened from five days to two days because of a failed fuel cell, the crew accomplished more than 90 percent of the objectives set for STS-2 before returning to a Rogers Dry Lake landing at Edwards. Major test objectives included the first tests in space of the fifty-foot remote manipulator arm. Also, twenty-nine flight test maneuvers were performed during the entry profile at speeds from Mach 24 (18,000 mph) to subsonic. These maneuvers were designed to extract aerodynamic and aerothermodynamic data during hypersonic entry into the earth's atmosphere. I always wondered whether Engle had surprised Dick Truly on reentry with high Gs as he did me sixteen years earlier in a T-38 at Edwards.

On his next mission, Engle was commander of STS-51I, which launched from Kennedy Space Center on August 27, 1985. The mission was acknowledged as the most successful space shuttle flight flown at that time. The crew deployed three communication satellites, the Navy SYNCOM IV-4, the Australian AUSSAT, and the American Satellite Company's ASC-1. The crew also performed the successful in-orbit rendezvous and repair of the ailing 15,000-pound SYNCOM IV-3 satellite. This repair activity saw the first manual grapple and manual deployment of a satellite by a crew member. STS-51I completed 112 orbits of the earth before again landing at Edwards on September 3, 1985. With the completion of this flight, Engle had logged over 224 hours in space.

Since Chuck Yeager retired from flying military jet aircraft, Engle has taken over the duty of opening the annual Edwards Air Show with a supersonic boom from an F-15 Eagle fighter.[8]

His height prevented Maj. **Frank Liethen** from being selected for the Air Force Manned Orbiting Laboratory program, as had happened at NASA a year or so earlier. The last chance to get into a space program and become an astronaut passed him by.

In early 1966 Liethen applied for a position as a demonstration pilot with the Thunderbirds. He made the initial cut and was one of four finalists who traveled with the team for some time. Team members made the final selection of new pilots. The team made their selection after many evaluation flights—Liethen was their choice. He was chosen to be the squadron executive officer for two years and then progressed to flight lead and squadron commander. Frank Liethen, the Dancing Bear, was definitely a leader.

One of the Thunderbird aircraft was a tandem-seated, dual-controlled model called the F-100F. Liethen had the opportunity to fly in the backseat with each of the pilots who specialized in a particular diamond-formation position. These flights would familiarize him with each of the positions where his wingman would eventually fly formation on him. Leading a demonstration flight is much more difficult than flying wing. The leader makes all the UHF radio commands for the flight and to the control tower. In addition, to make the formation appear smooth to the show crowd, the leader must be very smooth and precise on his flight controls so as to perform all of the maneuvers perfectly.

On October 12, 1966, Liethen was scheduled to fly the backseat of the F-100F with Capt. Robert H. "Bob" Morgan as the lead solo. Bob Morgan was four years younger than Liethen and had been flying fighters in Tactical

Air Command since he joined the Air Force eleven years earlier. He had completed one full year with the Thunderbirds and was designated to be the lead solo for his upcoming last year.

The regular air show season had finished for 1966. The team was now practicing from their home base at Nellis Air Force Base for the upcoming year. In the interests of safety, the practice air-show maneuvers were conducted several thousand feet higher than normal until the team got the wrinkles ironed out. All practice missions were graded from the ground. Since the last two aerial demonstrations had been critiqued as excellent, no particular improvement areas were discussed during the team preflight briefing.

The six Thunderbird pilots took off from Nellis and proceeded to their practice airfield at Indian Springs Air Force Auxiliary Field, Nevada (fifty miles northwest of Nellis). Captain Morgan, with Frank Liethen in the backseat, used the call sign Thunderbird 5, and Captain Bechel, the second solo, used Thunderbird 6. Arriving at the Indian Springs Auxiliary Field area, the team found the weather to be excellent for a practice show. The visibility was unlimited and the temperature was in the high 60s with light surface winds.

The six aircraft completed thirteen maneuvers without difficulty. The next maneuver was called "the solo opposing one-half Cuban Eight." To perform the maneuver, the two solo pilots approached each other straight and level at 480 mph and with slight lateral separation. They each called out on the UHF radio that they were twenty seconds from crossover. Just before passing each other at show center, they pulled up so that they passed or crossed with both aircraft climbing 30 degrees in opposite directions. Five Gs were used for the initial pull-up, but they gradually decreased as airspeed decreased in the climb. At the vertical or 90-degree point, both pilots tilted their heads back and looked for the other aircraft out of the top of their canopies. Azimuth corrections were started at this point so that one aircraft would pass over the other with a slight off-center lateral separation. The lead solo then instructed the second solo pilot to "pull it in," "float it," or "hold what you have" to allow both aircraft to pass close in the vertical plane with parallel flight paths. After passing, each aircraft continued downward to a 45-degree dive, then performed a one-half aileron roll to the upright position and recovered at minimum altitude. A plume of heavy white smoke trailed continuously from each aircraft to visually display their flight path, and at the top of the Cuban Eight the smoke trails appeared to merge as one passed over the other.

During the performance of the maneuver, the second solo was slightly late at his twenty-second checkpoint and increased his airspeed to 510 mph. The initial crossing of the two solos was slightly right of show center. The pull-up before the cross of the two solos was slightly late, and, therefore, the horizontal distance between the two solos when they attained the vertical position was greater than desired. The lead solo was higher than the second solo at the top (inverted position) of the half Cuban Eight, thereby lower in airspeed. Captain Morgan, the lead solo, told Captain Bechel, the second solo, "to float it" back toward him. The smoke trail on the lead solo's aircraft ceased after passing the top of the maneuver.

As the two aircraft passed, the base of Captain Morgan's front windscreen contacted the center of Captain Bechel's right horizontal stabilizer, thereby ripping three-fourths of the stabilizer off as it sliced through the entire length of Captain Morgan's canopy. The tip of Captain Morgan's vertical tail contacted Captain Bechel's right wing, slicing eight inches off the vertical tail and damaging the underside of Captain Bechel's wing.

Captain Bechel's aircraft yawed violently, but he recovered from a steep dive. He experienced partial failure of his flight controls and requested that another Thunderbird join him in close formation so as to report damage to his aircraft. Captain Bechel made an emergency landing at Indian Springs Auxiliary Field with damaged flaps and a failed antiskid braking system. Unable to stop on the runway, he lowered his tailhook and made a successful barrier engagement. He was uninjured.

Captain Morgan's aircraft did a half roll after contact with the other F-100, and the damaged front seat with the fatally wounded pilot fell away from the aircraft while streaming part of the chute from a rip in the side of the parachute case.

With Maj. Frank Liethen fatally injured in the backseat, the aircraft rolled once more and hit the ground at a steep angle and a high velocity with the afterburner still engaged. The U.S. Air Force had lost a great leader.[9]

After NASA test pilot **Bruce Peterson** crashed the M2-F2 on May 10, 1967, he was hospitalized and did not come back to work for eighteen months. He returned to the Dryden Flight Research Center as a project engineer. After obtaining an FAA medical waiver he began flying on a restricted basis in the Convair CV-990 and F-111. Eventually he became the director of safety, reliability, and quality assurance before retiring from NASA in 1981. He also continued to fly as a U.S. Marine reservist, retiring as a colonel in 1987. He then joined Northrop as a system safety engineer on the B-2 bomber for eleven years.

Portions of Peterson's spectacular crash landing in the M2-F2 were used in the popular weekly television series *The Six Million Dollar Man* during the opening credits of every episode that aired from 1973 to 1978. The background story of the show was the crash of fictional astronaut Steve Austin (played by Lee Majors) in a lifting body, shown in the opening credits of the show with NASA footage of Peterson's 1967 real-life accident. The accident scene was horrific; anyone watching it would assume that the pilot was killed. In the TV series Steve Austin was severely injured in the crash and "rebuilt" in an operation costing six million dollars—thus the show's title. His right arm, both legs, and his left eye were replaced by bionic implants that enhanced his strength, speed, and vision far above the human norm. Austin used his enhanced abilities to work for the Office of Scientific Intelligence as a secret agent (and as a guinea pig for bionics). Peterson complained that he disliked having his accident repeatedly played on television. He hated reliving his accident week after week as he watched the show, courtesy of Steve Austin. Bruce Peterson died of natural causes in Orange County, California, in May 2006 at the age of seventy-two.[10]

In the summer of 1966, Maj. **Chuck Rosburg** departed Edwards for combat crew training in the F-4 Phantom. After six months of training, he was combat ready for the war in Vietnam. He joined the 558th Tactical Fighter Squadron located at Cam Ranh Bay Air Base in South Vietnam. Rosburg flew 197 combat missions over Vietnam, about 400 hours flying time, earning the Distinguished Flying Cross and Air Medal. He had finally become the aggressive fighter pilot we had seen at parties and on the handball court.

In early April 1968, just as I headed off to war, Rosburg completed his one-year tour in Vietnam and returned to Edwards. He was assigned to the Vertical/Short Takeoff and Landing (V/STOL) Branch at Test Operations. After all his years flying the B-47 in SAC, flying the U-2 at Edwards, and combat experience in the F-4 in Vietnam, Rosburg was trusted to test one of the most difficult and dangerous types of aircraft. He learned to fly the British Hawker Siddeley XV-6B Kestrel, an early design of an aircraft that would eventually be called the AV-8 Harrier in the U.S. Marine Corps.

In late January 1969 Rosburg became a member of a test team tasked for an eleven-day period to evaluate the latest version of the Harrier, the GR Mark 1. The evaluation was conducted at the manufacturer's plant next to Dunsfold Aerodrome, Dunsfold, Surrey, England.

With Chuck Rosburg in the cockpit, the last test flight of the evaluation was scheduled for January 27, 1969. He lifted off vertically and started accelerating into forward flight. To prevent flying over a parked fire engine—pilots never fly over a manned vehicle because of the added danger—Rosburg applied rudder displacement and started to sideslip. An uncontrolled roll developed and either Rosburg applied the wrong rudder to correct this condition or he simply decided the situation was beyond his control—we will never know. The

Harrier flew into an uncontrollable bank. The aircraft was in almost 90 degrees of bank when Rosburg pulled the ejection handle. He blasted out of the cockpit sideways, and the rocket seat slammed him straight into the runway. The Harrier itself slid into the ground a short distance away. A movie cameraman captured Chuck Rosburg's death at twenty-four frames per second.

Rosburg was the victim of intake momentum drag yaw, the same phenomenon that Bill Bedford (chief test pilot for Hawker Siddeley and first pilot to fly the Harrier) had encountered from the earliest days. There was a danger at speeds in excess of fifty knots in allowing the aircraft to point anywhere except the direction it was going. At low speeds it didn't have to point its nose where it was going. It could go backward or sideways quite safely. But at slightly higher speeds, before the fin of the airplane had sufficient airflow over it to control the airplane's direction, it was directionally unstable. The air rushing into the Pegasus intakes, ahead of the center of gravity, made the nose want to turn sideways even more. The intakes were fighting the fin until the fin got more powerful as speed increased. So, between 50 and 110 knots, the amount of safe sideways movement was small.

All that Rosburg had to indicate sideways movement was a simple little wind vane mounted on a post outside the windscreen. If the vane was pointing roughly ahead, then the pilot knew the Harrier was going forward. If it was pointing off to the right or left, it indicated how much sideways movement he had. If the Harrier had been flying backward, then the vane would be pointing straight at him. Although the vane would tell the pilot that he was going sideways at 90 degrees, it did not tell him whether it was safe or not to do so. Airspeed was the critical factor, and the pilot got that only by looking at the airspeed indicator.

British test pilot John Farley had said long before that it was only a matter of time before someone failed to realize quickly enough that the airspeed was too high for the position of his wind vane. Unfortunately, Chuck Rosburg had proved the truth of that prediction.[11]

The summer after Rosburg's accident, the commanding general of Edwards Air Force Base unveiled a memorial that would be on permanent display in the newly renamed base gym: Rosburg Gym.

The general stated in his talk, "Our purpose here today is not to dwell in the past but bring tribute to one of our own test pilots who made great contributions to his profession, a man who was a champion both in the air and on the ground."[12]

After nearly ten years of test flying at Edwards, Maj. **Pete Knight** departed for Vietnam in 1968, where he completed 253 sorties in the F-100. Following his combat tour, he served as test director for the F-15 System Program Office at Wright-Patterson Air Force Base, Ohio. In this capacity, he became the tenth pilot to fly the F-15 Eagle and completed some of the initial evaluations of the fighter. Following an assignment as director of the Fighter Attack System Program Office, he returned to Edwards as vice commander in 1979, his last active duty assignment. During his tour Knight remained an active test pilot in the F-16 Combined Test Force.

Knight's greatest accomplishment was the major aviation milestone he set in 1967, when he piloted the modified X-15 number two to a speed of 4,520 mph (Mach 6.7), a speed that to this day is the highest ever attained in an airplane. After thirty-two years of service and about seven thousand hours in the cockpits of more than one hundred different aircraft, he retired from the Air Force in 1982 as a full colonel.

In 1984 Knight was elected to the city council of Palmdale, California, and four years later became the city's first elected mayor. Then he was elected to serve in the California Assembly representing the 36th District. When that four-year term expired, he was elected as a state senator representing California's 17th Senate District. Knight achieved statewide attention when he authored Proposition 22, the voter-approved Defense of Marriage Act that detractors dubbed the "Knight Initiative," which defined marriage in California as the union of one man and one woman.

Knight discussed his ongoing efforts to defend Proposition 22 during the taping of a Santa Clarita Valley Press Club *Newsmaker of the Week* interview:

> The idea of changing the definition of marriage is kind of foreign to me. That's what it's all about.
>
> Marriage is defined. Marriage is a natural evolution of man and woman. Man and woman together make a marriage. That's not discrimination.
>
> You know, they [critics] talk about equal rights, but there is no right to marriage. There is no civil right that says that you should be allowed to marry a man and a man.[13]

David Knight was Pete Knight's only son to follow his father into a career as a military pilot. David graduated from the Air Force Academy and flew jets in the first Persian Gulf War.

"My father was certainly proud of me," David Knight recalled. "He spoke at my pilot training graduation."

David, now forty-two and a custom furniture maker in Baltimore, was married in San Francisco to his longtime partner, Joseph Lazarro, an architectural designer, two days before the California Supreme Court stopped same-sex weddings in the state. San Francisco had tested state law by sanctioning almost four thousand gay and lesbian marriages. The elder Knight declined to talk to the press about his son's legally uncertain marriage, though he insisted his drive to keep gays from marrying didn't make him antigay.

Pete Knight died on May 7, 2004, at seventy-four, at City of Hope National Medical Center in Duarte, California, where he had been receiving treatment for leukemia.

North American test pilot **Van Shepard** flew the XB-70 Valkyrie forty-six times, twenty-three from the left seat and twenty-three from the right. After the program was canceled and the remaining XB-70 was ferried to the Air Force Museum at Wright-Patterson, Shepard joined Aero Spacelines of Santa Barbara, California, in 1968. By then Aero Spacelines was designing aircraft that was a big improvement over the Super Guppy that had made an emergency landing at Edwards in 1965. The new plane was a Boeing 337 airframe, structurally modified with turbine engines and referred to as the Mini Guppy Turbine (MGT). With an FAA registration of N111AS, the plane first flew on March 13, 1970, making it the first of the new generation of Guppies to fly. The 377MGT was originally designed for the commercial market, to haul ferryboats, helicopters, and industrial machinery with its heavier load-carrying capability.

The 377MGT was involved in an accident at Edwards on May 12, 1970, with Van Shepard piloting. The aircraft was the flight test vehicle being utilized for certification in accordance with Federal Air Regulations Part 25. At the time of the accident, 377MGT was engaged in one of a number of flight tests intended to establish the aircraft's optimum takeoff rotation speed with the wing flaps set at 20 degrees and the critical engine inoperative.

The accident occurred during the sixth takeoff following the scheduled cut of the number-one engine at an indicated airspeed of about 109 knots. The takeoff was made on Runway 22, and the wind was from approximately 200 degrees at about 10 knots. Rotation occurred at 114 knots, and several seconds after rotation, according to one witness, the aircraft turned and rolled to the left, settling as it did so. The left wingtip subsequently contacted the ground, which resulted in the aircraft being forcibly yawed from an initial magnetic heading of about 245 degrees (according to the flight data recorder) to a final heading measured as about 20 degrees. As a result of this cartwheeling action, the forward section to the aircraft was rammed into the ground and demolished, killing the four crew members.

Dead in the accident was pilot Van Shepard, copilot Hal Hanson, flight engineer Travis Hodges, and flight test engineer Warren "Sam" Walker.

After flying the dangerous XB-70 to Mach 3 and 70,000 feet, Shepard was killed in a crash at just over 100 knots.[14]

Maj. **Doug Benefield** went on to flight test all the big ones at Edwards, including the Boeing B-47 Stratojet, KC-135 Stratotanker, and civilian 707, Douglas DC-8, Lockheed C-130 Hercules, and C-141 Starlifter aircraft.

In February 1967 Benefield attended an F-4 Phantom combat crew training squadron at MacDill Air Force Base, Florida. He then went to Vietnam for a year flying 174 combat missions from Cam Ranh Bay, South Vietnam, with the 12th Tactical Fighter Wing. He was shot down on one mission and ejected over the water, joining the Tonkin Gulf Yacht Club. I always wondered how the big man ever fit in that Phantom cockpit.

In 1969 the FAA requested that Benefield be transferred to their headquarters to work on development of safety and aircraft handling qualities for certification of the U.S. Supersonic Transport (SST) Program. While working with the British and French on joint certification requirements for the SST, Benefield got a chance to fly the ultimate fast jet transport, the Mach 2 Concorde.

Doug Benefield retired from the Air Force and joined the B-1 Division of Rockwell International as a test pilot. Soon he became the test program's chief test pilot and logged more time in the aircraft than any other pilot.

On October 2, 1981, President Ronald Reagan asked Rockwell International to develop a new version of the B-1 Lancer that would retain the original's general appearance while incorporating the state-of-the-art avionics and stealth technology that had evolved during the ten years since the original B-1 design had been frozen. The new aircraft would be designated B-1B, while the original four B-1 aircraft were redesignated B-1A.

In 1982 Rockwell was awarded a full-scale development contract to build the B-1B. During the summer, work started on the second B-1A prototype to modify it for the B-1B program. It was originally anticipated that this work would take twenty-one months, but in only nine months it was ready to fly. Its test functions were to cover stability and control, flutter, and weapons release; and although it was widely hailed by the press as the first B-1B, it was actually a modified B-1A. The flight test program depended on a single aircraft for a prolonged period, but at least it was a flight-proven aircraft.

The number four–built B-1A prototype, painted in three-tone desert camouflage markings, paid a visit to the 1982 Farnborough International Air Show in England, where it was the star attraction. Benefield guided the aircraft throughout the tour and was in his element.

Doug Benefield was elected president of the Society of Experimental Test Pilots in 1983. The society is made up of about two thousand test pilots from all over the world. They meet annually in the Los Angeles area to discuss recent developments in flight testing. The previous two years I had served as legal officer and treasurer for the society, and in 1983 I was the Central Section representative. Board meetings were held every month in Lancaster, California. Since the chairman of the Central Section could not attend, I represented him and all the other members of his section, which covered Kansas, Nebraska, South Dakota, North Dakota, Minnesota, Wisconsin, Illinois, Iowa, and Missouri. I worked closely with Benefield for a year on the Board of Directors. He had not lost his touch—he handled the board with the same precision I had observed in him flying the C-133 many years earlier.

The white number two B-1A prototype joined the B-1B test program in March 1983, with a huge B-1B painted on its tail in red, white, and blue. The plane that had been flown at Mach 2.22 five years earlier logged 261 hours in 66 flights over the next seventeen months in the B-1B test program.

On August 28, 1984, the number two prototype was just four flights away from concluding its allotted program. That day Doug Benefield was designated as an instructor pilot, with Maj. Richard Reynolds, USAF, and Capt. Otto Waniczek, USAF, as crew members, and scheduled for a test flight.

The original flight plan called for early morning takeoff and flight tests for four hours and twenty minutes. A performance takeoff was planned, followed by airspeed calibration tests in the Edwards tower flyby pattern, then static and dynamic minimum speed control tests in the Cords Road area. After refueling in flight, the aircraft would enter the Edwards Precision Impact Range Area, where it was to make two weapons-release passes, the first simulated and the second releasing five MK 82 high-drag bombs whose warheads were filled with concrete.

The mission was planned to end with touch-and-go landings back at Edwards with Benefield acting as a teacher to the command pilot, Major Reynolds. Sixty-nine days had elapsed since Major Reynolds had made a landing in the B-1, whereas to remain current in type, a landing was required every sixty days. The final test point was a landing touchdown load test using maximum braking effectiveness.

The engines were started and the big bomber began to taxi out. During the taxiing phase ground load survey turning tests were carried out on the ramp. After that test the aircraft proceeded to Runway 22 where a final preflight check revealed that scuffing during the turning tests had caused excessive wear to both nose-gear tires and two tires on the main gear. The aircraft then returned to the ramp for remedial work.

Edwards is one of the busiest military airfields in the world. The delay meant that the sequence of the mission had to be rescheduled so as to avoid disruption of other activities planned for that day. The revised test sequence now became: takeoff, weapons release, airspeed calibration, minimum speed control tests, and a landing. The first two phases were uneventful, though the airspeed calibration tests were curtailed after two passes due to thermal turbulence that threatened to make the data obtained invalid.

Some forty-three minutes after takeoff, the B-1B, accompanied by its F-111 chase aircraft, climbed to 6,000 feet in the Cords Road area and was configured for static minimum-control speed points, with wing sweep at 55 degrees; flaps, slats, and gear retracted; and the center of gravity at 45 percent of mean aerodynamic chord (MAC). This test was successfully carried out at a speed of about 250 knots. Normally, the center of gravity changes would be carried out automatically, but the nature of this test series called for manual operation.

The next test was dynamic minimum control speed, with the wings fully extended at 15 degrees, flaps, slats, and landing gear all down, the center of gravity at the aft limit for that configuration of 21 percent MAC, and a speed of 136 knots. The B-1B accelerated to 300 knots, whereupon sweeping the wings forward commenced, a process that took forty-six seconds. The flaps, slats, and landing gear were all extended, but the center of gravity was left unchanged unintentionally, far outside the limit for that particular configuration. As the airspeed decayed through 145 knots with the angle of incidence rising to 8.5 degrees, the aircraft suffered an uncontrollable pitch-up to an angle of 70 degrees. Despite the application of full forward stick and afterburner, the aircraft was in an irrecoverable position, and Major Reynolds initiated ejection.

The crew punched out of the crippled aircraft in the crew-ejection capsule, a system that involved a twenty-one-foot segment of the fuselage containing the entire cockpit area being blasted out of the plane and carried to the ground by three parachutes. The system had been part of only the first three B-1As and was replaced by four individual McDonnell Douglas aircraft escape system ejection seats (ACES) aboard the fourth B-1A and all production B-1Bs. Even so, all should have been well for the crew. The crew-escape module had been designed to separate at low altitudes and adverse attitudes, correct the attitude and regain sufficient height for the parachutes to deploy, and then make a soft landing on the impact bladders.

The parachutes did deploy satisfactorily. However, one of the explosive repositioning bolts failed to function, and the module made a hard nose-down landing. The huge capsule slammed into the desert floor just a few feet from the fiercely burning wreckage of the bomber. Major Reynolds and Captain Waniczek were injured but survived. Doug Benefield was killed.[15]

In a *Life* magazine interview earlier that year, Benefield sounded almost poetic: "We're flying every day because we're kids who never stopped dreaming. We're always looking at another frontier, whiskering up to the edge of human and technological endurance."

Benefield also told the the local paper in 1984, "On the average day, you aren't flying at all. Instead you are going to meetings and studying. You're always studying."

"Some guys like to fish," he added. "Me, I just love to fly."[16]

Pancho Barnes was stricken with cancer in the early 1970s. Although the redoubtable woman vowed never to surrender and went on to survive two operations, the old zest for life gradually faded. Grover "Ted" Tate, a General Dynamics flight representative whom I had met on the F-111A program at Edwards, was a close friend of Pancho and found her in her home about a week after she died, alone, in 1975. Her son, Bill, and Ted planned to scatter her ashes where the Happy Bottom Riding Club had been located. They got approval from the commander of the Air Force Flight Test Center, Bob Rushworth. The pair took off from a nearby airfield and dropped the urn of ashes. The ashes were the same color as the Mojave Desert.

Bill Barnes became a pilot and owned a flying business at nearby Fox Field in Lancaster. He died in October 1980, crashing his P-51 Mustang a few miles west of the site of Pancho's old ranch while on his way to an air show at Edwards.

Of Pancho's glamorous era, little now remains: some concrete foundations and the remains of a fanciful stone fountain near the Edwards Gun Club and firing range. The entrance road to the range has been named Adams Way, in honor of Maj. Mike Adams, my friend who was killed in the X-15 in 1967.

The dim, rectangular outline of a dirt airstrip can still be made out from the air. There is a battered door from the ranch pickup, still faintly lettered, resting against a wall in the Air Force Flight Test Center Museum. Pancho Barnes stories still circulate freely in the flight community, some titillating, most nostalgic, all now recounted with tolerant smiles. For many years now, the people at Edwards have gathered on the site of the Happy Bottom Riding Club for an annual barbecue that goes far into the night. And, in a hangar in nearby Mojave, Pancho's black-and-red Travel Air Mystery Ship is gradually being returned to its original splendor.

As always, Pancho had the last word when she told Chuck Yeager, "Well f— it, we had more fun in a week than most of the weenies in the world have in a lifetime."[17]

Colonel **Chuck Yeager** left the test pilot school in July 1966 and assumed command of the 405th Fighter Wing at Clark Air Base in the Philippines. During this tour he flew the B-57 and F-100 for 127 combat missions over Vietnam. In February 1968 he took command of the 4th Tactical Fighter Wing at Seymour Johnson Air Force Base, North Carolina, which was deployed to Korea during the *Pueblo* crisis. He then was promoted to brigadier general and became vice commander of the 17th Air Force at Ramstein Air Base, Germany, and later was assigned as U.S. defense representative to Pakistan.

Yeager commenced his final active duty assignment as director of the Air Force Safety and Inspection Center at Norton Air Force Base, California. This was an ironic assignment for a pilot who had damaged a twin-engine B-57 jet bomber at Edwards by crashing into a bus that was being used as North American Aviation's ground data station for the XB-70 program. The bus was routinely parked in a space parallel to a taxiway. Yeager damaged the bus with an extended-wing model of the B-57 as he was taxiing. With little ado it was decided that there was no pilot error—it was the fault of the contractor for parking the bus in plain sight, where it had been for months. It is doubtful if any other Edwards pilot would have been quietly excused, but Yeager managed to blame the incident on the bus driver.

After a thirty-four-year military career, Yeager retired on March 1, 1975. At the time of his retirement, he had flown about 9,000 hours in more than 330 different types and models of aircraft.

A couple of years ago Chuck Yeager and I formed a stand-up comedy team. For a few moments we were good enough to be on the Jay Leno or David Letterman TV shows. We had the crowd in our hands. Yeager knows a few one-liners and how to get a laugh. For a while I was his straight man, and then we exchanged places. We were at the top of our game—we owned the stage, to use a theater expression. So how did this happen?

Yeager and I are members of the Society of Experimental Test Pilots and were attending the 2003 fall symposium, an annual event, in Los Angeles. During the Friday evening reception, Dr. James O. "Jim" Young, chief historian of Edwards, and I were walking through the crowd looking to get our glasses refilled. Yeager was holding court at a large table with ten to twelve people hanging on his every word. The young Air Force officers sitting at his table were all military pilots presently attending Test Pilot School; their attractive wives were sitting next to them. All eyes were glued on the grand old man of flight test. Yeager spotted Young, who had known and worked with Yeager for over twenty years, and signaled him over to his table.

Young introduced me to Yeager, saying, "Chuck, this is George Marrett. He wrote a story about the Lockheed NF-104 rocket-powered Starfighter you flew years ago."

"The article was published in *Air & Space* magazine," he added.

I had graduated from the test pilot school thirty-nine years earlier when Yeager was the commandant of the school. Hundreds of pilots went through the school while Yeager was stationed there. I was just a lowly student, a face in the crowd, and didn't expect him to remember me. However, Yeager did remember my story and complimented me.

"How many of the pilots I trained at the school went into space?" Yeager asked his assembled court.

No one responded, so Yeager answered his own question. "Twenty-six became astronauts!" he emphasized.

It was an impressive number, but everything about Yeager was impressive.

I told Yeager and his court that besides the magazine article, I had also written a book titled *Cheating Death: Combat Air Rescues in Vietnam and Laos*, which had been published six months earlier by the Smithsonian Institution Press. Over sixteen thousand copies of a print run of twenty thousand had been sold.

"Do you know how many copies of my book titled *Yeager* have sold?" Yeager asked, as he looked me directly in the eye. "Three and a half million," he said with a wide grin.

"I hope you made a few bucks on it," I answered. "Mine is really dropping in price. It was first offered for sale last winter. Last January it could be purchased at Barnes & Noble at a 20 percent discount, and in February Amazon sold it for 30 percent off."

I told Yeager that the price of my book was dropping faster than the stock market and it would probably be remaindered at Wal-Mart for eighty-nine cents by Christmas. Yeager laughed at my remark and the students and their wives got a good chuckle.

I continued, after the laughter subsided, "Chuck, Smithsonian Institution Press sent my book to you last year asking for a review or dustcover quote, and they never got an answer."

Yeager responded that he was a year behind in reading all the books sent to him for review. He said at his age he didn't read much anymore.

His comments were the perfect set up for me. I agreed with Yeager and told him and his assembled court I didn't read much either. Matter of fact, my wife told me when I was first offered a book contract, "You might be the only person to write a book that has never read one!"

That caustic remark put a big smile on Yeager's face.

Yeager shook his head in agreement and told the crowd that a huge number of people send books to him to autograph. He explained that he always degrades the book by writing both the name of the person who wants the autograph and then signing it. That way the book doesn't have as much value if they choose to resell it.

I concurred with Yeager on that point and indicated that I also inscribe my books to a particular person before I sign it. In addition I watch the Internet to see if anyone is selling a book on an auction site that I have autographed. I did buy one of my autographed books on eBay, just to find out who sold a copy I gave to them for free.

Yeager and I were on a roll. I had given enough book talks to develop a good awareness of audience reaction to my own put-downs, and Yeager can hold his own in any crowd. It would sure be fun to spar with Yeager and take our show on the road, but I don't have a publicity agent.

I'll bet Yeager does.

On my return from the war in Southeast Asia in May 1969, I joined Hughes Aircraft Company as an experimental test pilot and lived in Woodland Hills, California. For twenty years I flew test programs in the Douglas TA-3B Skywarrior, Grumman F-111B Sea Pig, and F-14A Tomcat, developing the AWG-9 Radar and AIM-54 Phoenix missile for the U.S. Navy. I flew the Douglas WB-66D Destroyer, developing the F-15 Eagle APG-63 attack radar for the Air Force. I flew the North American T-39D Sabreliner, developing the F-18 Hornet APG-65 Radar for the U.S. Navy. I also flew the McDonnell F-4D/E Phantom, developing the AGM-65 Maverick missile, the McDonnell F-4D Phantom, developing the F-16 attack radar, and a Douglas A-3A Skywarrior on an early version of the B-2 Stealth Bomber radar. The book I wrote about my years as a test pilot for Hughes, developing electronic weapons for the Cold War, titled *Testing Death: Hughes Aircraft Test Pilots and Cold War Weaponry,* was published in 2006 by Praeger Security International.

In October 2004 the Naval Institute Press published my book *Howard Hughes: Aviator*. Several of the mechanics and engineers that worked on the military planes I flew while I was employed with Hughes Aircraft Company also worked on Hughes' personal aircraft. Similarly, several of the older test pilots, who had been with the company since the end of World War II, had flown with Hughes as a passenger or had flown as a copilot for him. For that book I reconnected with six people who introduced me to others; ultimately, I interviewed nine people about their unique flying experiences with Howard Hughes. Additionally, I interviewed another twelve persons who were either wives, sons, or daughters of men who had flown with Hughes. Others whom I got information from were people who had left writings and transcripts of their rich and colorful time with the mysterious aviator. All were pleased that a nonfiction account of Hughes' aviation achievements was being documented for future generations. Since many were quite elderly, my book is likely to be the last firsthand account of Howard Hughes, the pilot, based on personal interviews.

Soon after retiring from Hughes Aircraft Company in 1989, I helped found the Estrella Warbird Museum at the Paso Robles, California, airport, about twenty miles north of my home in Atascadero. During the last couple of years museum members have acquired a number of former military aircraft that I flew either in the Air Force or at Hughes Aircraft Company. Members refurbish these planes and put them on static display.

As an Air Force student pilot I soloed in a single-engine propeller Beechcraft T-34 aircraft in 1958. It was in this plane that I learned to do stalls, loops, and spins. I almost became airsick on one acrobatic flight but managed to recover and passed that important aviation hurdle. Our museum acquired a Navy T-34 that belonged to the Naval Air Station Lemoore Aero Club. A pilot flying the plane near the Pacific Ocean ran out of gas and landed gear-up in an open field. Unfortunately, the plane slid into a barbed wire fence and severely damaged the wing. This T-34 was given to us and is now being restored by museum members.

The first jet I flew was the Cessna T-37. After just thirty flying hours in the T-34, I moved up to piloting a twin-engine jet. For the first time I wore a jet helmet with attached oxygen mask and sat on an ejection seat. What a thrill it was to fly my first jet and finally become a "jet pilot." Museum members found a discarded Air Force T-37 in Indiana and trucked it all the way to our museum in California.

I earned silver Air Force wings flying the Lockheed T-33. It was in the T-33 that I first learned to fly on instruments, to fly in formation, and to fly at night. The T-33 was also the first jet I flew that was an offspring of a fighter aircraft, the F-80 Shooting Star. In was in the T-33 that I flew maintenance test missions, my first testing in an aircraft. I accumulated over two thousand hours in the plane flying it both in the Air Force and for Hughes Aircraft Company. Our museum members found a T-33 in Madera, California, that the Air Force had lent to the city. The plane had fallen into disrepair and the Air Force was talked into transferring the T-33 to the Estrella Warbird Museum.

I first flew supersonic in the North American F-86L Sabrejet. I made a vertical dive into the Okefenokee Swamp and boomed the alligators. The Naval Air Weapons Station China Lake, California, operations department obtained many F-86s back in the 1960s and 1970s to modify into drones. Museum members scrounged through mounds of surplus parts from five planes and assembled an F-86 that looks so authentic you would think it is still flyable.

I tested the F-4C Phantom and F-4E prototype at Edwards and later flew the F-4C, F-4D, and F-4E as a test pilot for Hughes Aircraft Company. Our museum found a Marine F-4S at Marine Corps Air Station Yuma, Arizona. The wings were torched off so that the huge fighter would fit on a flatbed truck and be transported to California. The wings were reassembled, the plane painted, and decals attached so that our F-4S appears in its original military markings.

At Edwards I had a chance to fly the F-5A Freedom Fighter at the end of its test program. Our museum picked up an F-5E that had been used as a "bad guy" in a Navy aggressor squadron at Naval Air Station Fallon, Nevada. The aircraft's performance matched that of the Russian Mikoyan MiG-21 aircraft, so the Navy used the F-5E to teach their pilots how to dogfight with the enemy. The Freedom Fighter excelled in that role by providing a small, fast, very agile, and hard-to-see adversary—especially since it was flown by very experienced pilots using the same tactics flown by potential adversaries that U.S. pilots were likely to face in actual combat. Our newly acquired F-5E is painted in two-tone blue and gray camouflage with a big red Russian star on the tail. Northrop produced 1,871 F-5s and a further 776 were built under license in Canada, Spain, Switzerland, Korea, and Taiwan. The F-5 served as a jet fighter in thirty foreign countries and was flown in combat in Vietnam. However, it didn't really affect the outcome of the Cold War nor did it have an illustrious combat history. After all the years have gone by, the T-38 and F-5 might best be remembered as the aircraft that could go vertical.

Test pilot friend Larry Curtis was killed and Joe Stroface severely injured in the crash of an OV-10A Bronco. OV-10As were discarded by the military years ago, and some were given to the California Department of Forestry for fire patrol. Our museum found a surplus Bronco in Sacramento. Like our other planes, the OV-10 was disassembled on the spot, trucked to our airport, and then reassembled. We continue to look for discarded military aircraft to restore and put on public display.

While searching for more military aircraft, I encountered the number seven–built Grumman F-111B. It was in the disposal area at Naval Air Weapons Station China Lake, California. The plane was the last F-111B built, the best of the lot, and the one I used as a Hughes Aircraft test pilot to fire many AIM-54A Phoenix air-to-air missiles at drones over the Pacific Ocean. Now the plane is simply a 70,000-pound hunk of metal slowly aging on the windswept high desert of California. My hope is that some aviation museum will claim this historic plane and give it a proper resting place.

Recently the Estrella Warbird Museum acquired an F-14 Tomcat from the Navy to put on static display. I flew Tomcats for ten years from Naval Air Station Point Mugu, California, as a Hughes test pilot developing the AWG-9 radar and firing AIM-9 Sidewinders, AIM-7 Sparrows, and AIM-54A Phoenix missiles at drones in the Pacific Missile Test Center (PMTC) overwater test range. F-14s have been removed from active naval service and have been replaced by an advanced version of the F-18 Hornet, the F-18E/F Super Hornet. This plane, with an improved Hughes Aircraft Company radar, will serve the Navy well into the twenty-first century. I witnessed the F-14 fly into the Paso Robles airport on its final flight about two years ago. While museum members cheered, I shed a tear, realizing I had seen my last Tomcat in the sky.

Sometimes on a warm summer evening before sunset I go out to the museum by myself and walk around our assembled group of about twenty restored military warbirds. I reflect back on the years long ago when I flew airplanes like the ones we now possess. I think about the pilots I flew with in Air Force flight school, in the 84th Fighter Interceptor Squadron, in combat in Vietnam, and in flight test at Edwards. Many have died, and I have lost track of most of the rest except for a few I knew in Test Pilot School and in combat. The years have flown by too fast, and my memory of time in the air is fading into the mist. But the assembled aircraft are a reminder for those who visit our museum now and in the future of the airborne warriors of the past who dedicated their lives and future to our country's defense during the Cold War.

For the last seven years I have flown a twin-engine Beech King Air turboprop aircraft for a local company that makes hangar doors. While on a trip to Georgia, I had the opportunity to visit the Air Force Museum of Aviation at Robins Air Force Base. F-104A Starfighter serial number 56-0817 was on display there. I remembered flying the plane at Edwards chasing the X-15, XB-70, and other test aircraft of that era. My recollection was that 56-0817 was one of the fastest birds in our fleet and pilots didn't have any emergencies with the single-seat fighter. I was pleased this particular Starfighter survived the hectic times at Edwards and is now on static display.

Another one of our Edwards F-104A Starfighters, serial number 56-0763, may set a speed record in the future. Several private individuals purchased the plane from a scrap yard in Maine and removed the wings and horizontal tail. The fuselage was painted in patriotic red, white, and blue colors and the vehicle named the *North American Eagle*. The owners plan to install a highly modified General Electric J79 engine and go for a land speed record over the Bonneville Salt Flats and the Black Rock Desert of Nevada. Their goal is to break the current world land speed record set October 15, 1997, when Royal Air Force pilot Andy Green drove the Thrust SSC to a new world record of 763 mph, or Mach 1.02. The car's owner, Richard Noble from England, held the previous record set in 1983 with Thrust2 at 633 mph. For over twenty-four years now the British have held the land speed record. Now this unique group of Americans, who are working on the *North American Eagle* much like the aviation pioneers of the Mojave Desert sixty-five years earlier, believes it is time to bring the land speed record back to the United States.

I had a dream as a student pilot in the Air Force to fly the sleek F-104 Century Series fighter. My dream came true when I was accepted into the Air Force Test Pilot School. As a test pilot I flew the Starfighter to Mach 2.15 and zoomed to 80,000 feet. I reached the edge of space and safely returned. The F-104 was one of the most streamlined aircraft ever built—and one of the most deadly and dangerous planes in the Air Force inventory during the Cold War. But the Starfighter was also a superb example of all the aircraft designed, tested, and flown to speed and altitude records at Edwards.

Several years ago the Estrella Warbird Museum did locate and restore a former NASA Lockheed F-104G Starfighter that had spent time at Edwards. Lockheed Aircraft Corporation in Burbank, California, built the plane in 1962 for the German Luftwaffe. Later it was transferred to NASA at Edwards where Chuck Yeager and several astronauts flew the plane. The aeronautical engineering program of California Polytechnic State University, San Luis Obispo, eventually obtained the aircraft and used it for student instruction and education for a few years. Finally, the plane was given to our museum and trucked to the Paso Robles airport. Now the F-104G, in bright blue-and-white NASA colors, sits in front of our hangar next to the entry road where visitors can't miss it. Our F-104G Starfighter is now the marquee at our museum, and my name is proudly displayed just below the canopy.

Flight testing at Edwards, and associating with world-class test pilots, has been the gift of a lifetime. I'm proud to say I have flown with the giants of aviation of my era. For sixty-four years I've chased after the little silver whistle since I first saw one as a child during World War II. The silver whistle was symbolic of my love for aviation and higher aerial achievement. Test flying and the defense of freedom have been my life and my passion—it's been an unforgettable ride.

Acronyms and Abbreviations

AAF	Army Air Forces, 1941–47
ACES	aircraft escape system ejection seats
ADF	automatic direction finding
AFFTC	(Edwards) Air Force Flight Test Center
AFIT	Air Force Institute of Technology
AFROTC	Air Force Reserve Officer Training Corps
ALT	approach-and-landing test
AO	airdrome officer
ARPS	Aerospace Research Pilot School
BOQ	bachelor officers' quarters
CAT	clear air turbulence
CBU	cluster bomb units
CG	center of gravity
CM	command module
E&E	(Air Force) Escape & Evasion School
EGT	exhaust gas temperature
FAA	Federal Aviation Administration
FAC	forward air controller
FAI	Federal Aeronautique Internationale
FCF	functional check flight
FIS	fighter interceptor squadron
G-force	the force of gravity or acceleration on a body
GCA	ground control approach
GD	General Dynamics
HiCAT	High Altitude Clear Air Turbulence program
ICBM	intercontinental ballistic missile
IFR	instrument flight rules
ILS	instrument landing system
JATO	jet-assisted takeoff
LAADS	Los Angeles Air Defense Sector
LARA	Light Armed Reconnaissance Aircraft program
LEM	lunar excursion module
LOX	liquid oxygen

LTV	Ling-Temco-Vought
MAC	mean aerodynamic chord
MAP	Military Assistance Program
MIDAS	Missile Detection and Alarm System
MOL	Manned Orbiting Laboratory
NAA	North American Aviation
NACA	National Advisory Committee for Aeronautics
NAS	naval air station
NASA	National Aeronautics and Space Administration
PIRA	(Edwards) Precision Impact Range Area
PMTC	Pacific Missile Test Center
PT	physical training
RCS	reaction control system
RIO	radar intercept officer
ROTC	Reserve Officer Training Corps
RTAFB	Royal Thailand Air Force Base
SAC	Strategic Air Command
SAM	surface-to-air missile
SPO	systems program office
SST	supersonic transport
TAC	Tactical Air Command
TACAN	Tactical Air Navigation system
TFR	terrain-following radar
TFS	tactical fighter squadron
TFX	tactical fighter, experimental
TPS	(Air Force) Test Pilot School
TWA	Trans World Airlines
TWX	military Teletype message
UFO	unidentified flying objects
UHF	ultrahigh frequency
USAFE	United States Air Forces in Europe
V/STOL	vertical/short takeoff and landing
VD	velocity dive
VFR	visual flight rules
VTOL	vertical takeoff and landing
WSO	weapon systems operator

Notes

Chapter 5. Cold War

1. U.S. Air Force Web site, "Biographies: Major General Robert M. White," http://www.af.mil/bios/bio.asp?bioID= 7574.
2. "Air Force Captain Found Dead of Gunshot," *Novato Advance*, March 28, 1963.

Chapter 6. Charm School

1. Robert Smith Web site, "NF104," http://www.nf104.com/stories/stories_13.html.
2. Batnet.com, "NF-104 Space Pilot Trainer," http://www.batnet.com/mfwright/NF104/html.

Chapter 7. Space Cadets

1. Material on Pancho Barnes is from the following sources: *The Happy Bottom Riding Club*, by Lauren Kessler (New York: Random House, 2000); *Pancho: The Biography of Florence Lowe Barnes*, by Barbara Heather Schultz (Lancaster, Calif.: Little Buttes Publishing, 1996); *The Lady Who Tamed Pegasus: The Story of Pancho Barnes*, by Grover Ted Tate (Bend, Ore.: Maverick Publications, 1986).
2. Tate, *Lady Who Tamed Pegasus*, 25.
3. This material on Pancho Barnes is from http://www.chuckyeager.com/specialfeatures/ Pancho3.htm.

Chapter 8. Flight Test Operations

1. "The Oxcart Story," *Air Force Magazine* 77, no. 11 (November 1994), http://www.afa.org/magazine/nov1994/1194oxcart.asp.

Chapter 11. Metal Falling from the Sky

1. "F-111Crash," *Antelope Valley Press,* January 22, 1967.
2. "Airman Medal Given Posthumously," *Desert Wings,* March 24, 1967.
3. Carl Posey, "A Sudden Loss of Altitude," *Air & Space*, June/July 1998; "Memorial Services Held for Robert H. Lawrence," *Desert Wings*, December 15, 1967.
4. Quoted in J. Alfred Phelps, *They Had A Dream* (San Rafael, Calif.: Presidio Press, 1994), 54.

Epilogue

1. "TWA Jetliner Crash," *Washington Post,* December 2, 1974.
2. Information on TWA Flight 514 is from Richard Witkin, *How and Why Did Flight 514 Kill 92 People* (New York: Arno Press, 1976).
3. Much of the information on Bill Anders is found on the NASA Web site, http://www.jsc.nasa.gov/Bios/htmlbios/anders-wa.html.
4. "Air Force Pilot Says Criticism of F-111 Not Justified by Facts," *Antelope Valley Ledger-Press*, March 20, 1971.
5. I found much of the information on Jim Hurt's F-111 accident from photocopied excerpts of a report sent to me at my request by John J. Clark Jr., Chief, Reports Division, Headquarters Air Force Safety Agency at Kirkland Air Force Base, New Mexico, on November 27, 1995.
6. Much of the information on Fred Haise is found on the NASA Web site, http://www.jsc.nasa.gov/Bios/htmlbios/haise-fw.html.
7. Much of the information on Jim Taylor's and Pierre du Bucq's T-38 accident came from photocopied excerpts of a report sent by John J. Clark Jr., Chief, Reports Division, Headquarters Air Force Safety Agency at Kirkland Air Force Base, New Mexico, on November 27, 1995 (see note 5 above).
8. Much of the information on Joe Engle is found on the NASA Web site, http://www.jsc.nasa.gov/Bios/htmlbios/engle-jh.html.
9. "T-Bird Crash Kills Two, One Was Past Local," *Desert Wings,* October 14, 1966.
10. Much of the information on Bruce Peterson is found on the NASA Web site, http://www.nasa.gov/centers/dryden/news/Biographies/Pilots/bd-dfrc-p012.html.
11. Information on Chuck Rosburg is from Bruce Myles, *Jump Jet: The Revolutionary V/STOL Fighter* (San Rafael, Calif.: Presidio Press, 1978).
12. "Base Names Gym to Honor Major Rosburg," *Desert Wings,* July 3, 1970.
13. Santa Clarita Valley Press Club, *Newsmaker of the Week,* http://www.scvhistory.com/scvhistory/signal/newsmaker/sg042504.htm.
14. "Four Die in EAFB Crash," *Antelope Valley Ledger-Gazette*, May 12, 1970; "Fatal Crash," *Los Angeles Times*, May 13, 1970.
15. Much of the information on Doug Benefield's B-1B accident came from photocopied excerpts of a report sent by John J. Clark Jr., Chief, Reports Division, Headquarters Air Force Safety Agency at Kirkland Air Force Base, New Mexico, on November 27, 1995 (see notes 5 and 7 above).
16. Laurie Becklund, "Comrades Eulogize Fallen B-1 Test Pilot," *Los Angeles Times,* September 1, 1984.
17. The information for Pancho Barnes is from Kessler, *Happy Bottom Riding Club;* Schultz, *Pancho;* and Tate, *Lady Who Tamed Pegasus.*

Resources

"AFFTC Pilot Killed in French VSTOL Crash." *Desert Wings*, September 10, 1965.

"Air Force Captain Found Dead of Gunshot." *Novato Advance*, March 28, 1963.

"Air Force Pilot Says Criticism of F-111 Not Justified by Facts." *Antelope Valley Ledger-Press*, March 20, 1971.

"Airman Medal Given Posthumously." *Desert Wings*, March 24, 1967.

Ball, John Jr. *Edwards: Flight Test Center of the U.S.A.F.* New York: Duel, Sloan and Pearce, 1962.

"Base Names Gym to Honor Major Rosburg." *Desert Wings*, July 3, 1970.

Batnet.com. "NF-104 Space Pilot Trainer," http://www.batnet.com/mfwright/NF104/html.

Becklund, Laurie. "Comrades Eulogize Fallen B-1 Test Pilot." *Los Angeles Times,* September 1, 1984.

Berliner, Don. "The Air Races Are Back!" *Air Progress*, February/March 1965.

"B-70, F-104 Collide over Desert; 2 Killed, 1 Injured." *Desert Wings*, June 10, 1966.

"Category II Test F-111A Arrives Here." *Desert Wings*, January 21, 1966.

"Crash of XV-5A Jet Kills Major Dave Tittle." *Desert Wings*, October 7, 1966.

Crickmore, Paul F. *Lockheed SR-71 Blackbird*. London: Osprey Publishing, 1986.

"Everybody Likes LARA Apparently Except . . . ???" *Air Progress*, August 1966.

"F-111Crash," *Antelope Valley Press,* January 22, 1967.

"Fatal Crash." *Los Angeles Times*, May 13, 1970.

"Four Die in EAFB Crash." *Antelope Valley Ledger-Gazette*, May 12, 1970.

Gilmore, Ken. "The Truth about the Amazing F-111 from the Men Who Fly Them." *Popular Science*, May 1968.

Hallion, Richard P. *Test Pilots: The Frontiersmen of Flight*. Washington, D.C.: Smithsonian Institution Press, 1988.

Harvey, Frank. "I Rode Our Hottest Jet." *Popular Science*, May 1962.

Infield, Glenn. "Test Pilot Who Tamed the Black Phantom Killer Jet." *Male*, July 1968.

Jackson, Paul. *Mirage.* Runnymede, U.K.: Ian Allan Printing, 1985.

Johansen, Herbert O. "F-5—Our Bantam Supersonic Jet." *Popular Science*, March 1966.
———. "600-m.p.h. Vertifan Jet Can Hover Like a Copter." *Popular Science*, September 1966.
Kessler, Lauren. *The Happy Bottom Riding Club*. New York: Random House, 2000.
Madison, Stu. "Tilt-Wing Transport." *Aerospace Safety*, December 1963.
"Major Michael J. Adams Lost in Local X-15 Crash." *Desert Wings*, November 17, 1967.
Mallick, Donald M. *The Smell of Kerosene*. Washington, D.C.: NASA, 2003.
Marrett, George J. "Chasing the XB-70 Valkyrie." *Flight Journal*, June 2000.
———. *Cheating Death: Combat Air Rescues in Vietnam and Laos*. Washington, D.C.: Smithsonian Institution Press, 2003.
———. "Defending the Golden Gate." *Wings*, August 1999.
———. "Don't Kill Yourself." *Aerospace Testing International*, September 2005.
———. *Evaluation of Longitudinal Control Feel System Modifications Proposed for USAF F/RF-4 Aircraft*. Edwards AFB, Calif.: Air Force Flight Test Center, December 1967.
———. *F-4C Category II Follow-On Stability and Control Tests*. Edwards AFB, Calif.: Air Force Flight Test Center, May 1968.
———. *F-4C Stability and Control Tests with a TAC Training Load*. Edwards AFB, Calif.: Air Force Flight Test Center, September 1967.
———. *Howard Hughes: Aviator*. Annapolis, Md.: Naval Institute Press, 2004.
———. "Mach Buster." *Sabre Jet Classics*, Fall 2000.
———. *Major Accident Report NF-104, SN 56-756*. Edwards AFB, Calif.: Air Force Flight Test Center, 1965.
———. "Northrop F-5A Freedom Fighter." *Wings*, December 2005.
———. "Sky High." *Air & Space*, November 2002.
———. "Space Cadets." *Wings*, June 2004.
———. *Testing Death: Hughes Aircraft Test Pilots and Cold War Weaponry*. Westport, Conn.: Praeger Security International, 2006.
———. "Voodoo Curse." *Flight Journal*, June 2003.
McIntosh, Robert H. *Major Accident Report F-104B, SN 56-3721*. Edwards AFB, Calif.: Air Force Flight Test Center, 1963.
"Memorial Services Held for Crash Victims." *Desert Wings*, April 30, 1971.
"Memorial Services Held for Maj. H. L. Brightwell." *Desert Wings*, January 27, 1967.
"Memorial Services Held for Robert H. Lawrence." *Desert Wings*, December 15, 1967.
Miles, Marvin. "Test Pilot Killed in Crash of Second Vertical-Rising Jet." *Los Angeles Times*, October 6, 1966.

Myles, Bruce. *Jump Jet: The Revolutionary V/STOL Fighter*. San Rafael, Calif.: Presidio Press, 1978.
"Ohio Crash of OV-10A Claims Life of FTC Pilot." *DesertWings*, February 3, 1967.
"The Oxcart Story," *Air Force Magazine* 77, no. 11 (November 1994), http://www.afa.org/magazine/nov1994/1194oxcart.asp.
NASA Web site. http://www.jsc.nasa.gov.
Pace, Steve. *Valkyrie: North American XB-70A*. Fallbrook, Calif.: Aero Publishers, 1984.
Phelps, J. Alfred. *They Had a Dream*. San Rafael, Calif.: Presidio Press, 1994.
Plattner, C. M. "Space Training Flights in NF-104A Near." *Aviation Week & Space Technology*, August 9, 1965.
Posey, Carl. "A Sudden Loss of Altitude." *Air & Space*, June/July 1998.
"Ryan Believes Future Remains Strong for XV-5A and Follow-On Vertifan Aircraft." *Ryan Aeronautical News Bureau*, October 7, 1966.
Reed, R. Dale. *Wingless Flight*. Washington, D.C.: NASA, 1997.
Reynolds, Gen. Dick. "B-1A 74-0159 Flight 2-127 August 29, 1984." *Society of Experimental Test Pilots: Cockpit*, October/November/December 2004.
Robert Smith Web site. "NF104," http://www.nf104.com/stories/stories_13.html.
"Ruptured Guppy." *Air Progress*, January 1966.
Santa Clarita Valley Press Club, *Newsmaker of the Week,* http://www.scvhistory.com/scvhistory/signal/newsmaker/sg042504.htm.Scholin, Allan. "Mr. Mac's Deadly Family of Phantoms." *Air Progress*, March 1968.
Schultz, Barbara Heather. *Pancho: The Biography of Florence Lowe Barnes*. Lancaster, Calif.: Little Buttes Publishing, 1996.
Scutts, Jerry. *Northrop F-5 / F-20*. London: Ian Allan Printing, 1986.
"Services Held for 2 Pilots." *Desert Wings*, September 11, 1970.
Sherman, Samuel M. "The Edwards Air Force Base Encounter." *Independent International Pictures*, 1995.
"Six F-111 Flights Accomplished in Single Day on Wet Runways." *Desert Wings*, July 1, 1966.
Smith, Lt. Col. P. G. "The Day the Super Guppy Blew Her Top." *AIR FORCE Magazine*, April 1971.
Spick, Mike. *Modern Fighting Aircraft: B-1B*. New York: Prentice-Hall, 1986.
Stimson, Thomas E. "The Controversial B-70: Military Monster or Supersonic Transport?" *Popular Mechanics*, June 1963.
Tate, Grover Ted. *Bombs Awry*. Bend, Ore.: Maverick Publications, 1985.
———. *The Lady Who Tamed Pegasus: The Story of Pancho Barnes*. Bend, Ore.: Maverick Publications, 1986.

"T-Bird Crash Kills Two, One Was Past Local." *Desert Wings*, October 14, 1966.

Thompson, Milton O. *At the Edge of Space*. Washington, D.C.: Smithsonian Institution Press, 1992.

Thompson, Milton O., and Curtis Peeples. *Flying without Wings*. Washington, D.C.: Smithsonian Institution Press, 1999.

Thompson, William Murry. *Republic Flight Test Pilot*. Charleston, S.C.: W. Thompson, 2001.

"TWA Jetliner Crash." *Washington Post,* December 2, 1974.

U.S. Air Force Web site. "Biographies: Major General Robert M. White," http://www.af.mil/bios/bio.asp?bioID= 7574.

West, Richard. "Two Pilots Die as XB-70, Chase Aircraft Collide." *Los Angeles Times*, June 9, 1966.

Witkin, Richard. *How and Why Did Flight 514 Kill 92 People*. New York: Arno Press, 1976.

"The World's First Space Trainer." *Airmen*, October 1963.

Yeager, Chuck, and Leo Janos. *Yeager: An Autobiography*. New York: Bantam Books, 1985.

Index

About the Author

George J. Marrett was born in Grand Island, Nebraska, and graduated from Iowa State College in Ames, Iowa, in 1957 with a BS in chemistry. Marrett entered the Air Force as a second lieutenant from the Reserve Officers Training Corps program. Graduating from pilot training in 1959, he was assigned to advanced flying school and flew the North American F-86L at Moody AFB, Georgia. After four years as an interceptor pilot flying the F-101B Voodoo, he was selected to attend the USAF Aerospace Research Pilot School at Edwards AFB, California, in 1964. Upon graduation he was transferred to the Fighter Test Branch of Flight Test Operations at Edwards and completed three years of flight-testing.

In 1968–69 Marrett flew 188 combat missions in the Douglas A-1 Skyraider as a Sandy rescue pilot in the 602nd Fighter Squadron (Commando) from Udorn and NKP Royal Thai Air Force Bases, Thailand. He was awarded the Distinguished Flying Cross with two oak leaf clusters and the Air Medal with eight oak leaf clusters. Upon returning from the Vietnam War he joined Hughes Aircraft Company and spent the next twenty years there as an experimental test pilot. He has flown over forty types of military aircraft, logging 9,500 hours of flying time.

Marrett is the author of three books: *Cheating Death: Combat Air Rescues in Vietnam and Laos* (Smithsonian Institution Press, 2003), *Howard Hughes: Aviator* (Naval Institute Press, 2004), and *Testing Death: Hughes Aircraft Test Pilots and Cold War Weaponry* (Praeger Security International, 2006). He has also written numerous magazine articles.

Marrett and Jan, his wife of fifty years, live in California. They have two sons and four grandchildren.

The Naval Institute Press is the book-publishing arm of the U.S. Naval Institute, a private, nonprofit, membership society for sea service professionals and others who share an interest in naval and maritime affairs. Established in 1873 at the U.S. Naval Academy in Annapolis, Maryland, where its offices remain today, the Naval Institute has members worldwide.

Members of the Naval Institute support the education programs of the society and receive the influential monthly magazine *Proceedings* or the colorful bimonthly magazine *Naval History* and discounts on fine nautical prints and on ship and aircraft photos. They also have access to the transcripts of the Institute's Oral History Program and get discounted admission to any of the Institute-sponsored seminars offered around the country.

The Naval Institute's book-publishing program, begun in 1898 with basic guides to naval practices, has broadened its scope to include books of more general interest. Now the Naval Institute Press publishes about seventy titles each year, ranging from how-to books on boating and navigation to battle histories, biographies, ship and aircraft guides, and novels. Institute members receive significant discounts on the Press's more than eight hundred books in print.

Full-time students are eligible for special half-price membership rates. Life memberships are also available.

For a free catalog describing Naval Institute Press books currently available, and for further information about joining the U.S. Naval Institute, please write to:

Member Services
U.S. Naval Institute
291 Wood Road
Annapolis, MD 21402-5034
Telephone: (800) 233-8764
Fax: (410) 571-1703
Web address: www.usni.org